SpringerBriefs in Electrical and Computer Engineering

Syed Faraz Hasan

Emerging Trends in Communication Networks

Springer

Syed Faraz Hasan
School of Engineering and Advanced
    Technology
Massey University
Palmerston North
New Zealand

ISSN 2191-8112 ISSN 2191-8120 (electronic)
ISBN 978-3-319-07388-0 ISBN 978-3-319-07389-7 (eBook)
DOI 10.1007/978-3-319-07389-7
Springer Cham Heidelberg New York Dordrecht London

Library of Congress Control Number: 2014941088

Printed on acid-free paper

Springer is part of Springer Science+Business Media (www.springer.com)

To the trio: Tatheer, Nida, and Khadeeja

Acknowledgments

I extend my gratitude to my father who taught me how to write. His efforts in helping me do well in education and academics are outstanding. I cannot thank him enough for his hard work and dedication. I must also thank my mother for her love and affection throughout my life. She has always been a firm and reliable support for me.

I am grateful to my students at Sungkyunkwan University, who helped me in fine tuning this manuscript. I frequently got support from the faculty members at Sungkyunkwan University, which is recognized with gratitude. I thank my teachers and friends at NED University of Engineering and Technology, and my Ph.D. advisors at University of Ulster. They helped me a lot in building a strong foundation for this work.

Contents

Figures

Abbreviations

3GPP	3rd Generation Partnership Project
AIN	All IP Networks
AMPS	Advanced Mobile Phone System
ARP	Address Resolution Protocol
BSC	Base Station Controller
CC	Component Carrier
CMS	CoMP Measurement Set
CoMP	Coordinated Multipoint
CP/SDN	Control Plane SDN
CQI	Channel Quality Indicator
CSI	Channel State Information
CUE	Cellular User Equipment
D2D	Device-to-Device
DAD	Duplicate Address Detection
DeNB	Donor evolved Node-B
DHCP	Dynamic Host Configuration Protocol
DNS	Domain Name Server
DNSSL	DNS Search List
DOS	Disk Operating System
DUE	D2D User Equipment
DUID	Device User ID
DV	Distance Vector
EUI	Extended Unique Identifier
FEC	Forward Error Correction
FHSS	Frequency Hopped Spread Spectrum
GENI	Global Environment for Networking Investigation
GSM	Global System for Mobile Communication
HARQ	Hybrid Automatic Repeat reQuest
HLR	Home Location Register
HSPA	High Speed Packet Access
I2RS	Interface to Routing System
IAID	Interface Association ID
IANA	Internet Assigned Numbers Authority
ICMP	Internet Control Messaging Protocol

IMT-A	International Mobile Telecommunications Advanced
IP	Internet Protocol
IT	Interference Tracing
ITU	International Telecommunications Union
LMDS	Local Multipoint Distributed Services
LTE	Long Term Evolution
M2M	Machine-to-Machine
MAC	Medium Access Control
MME	Mobility Management Entity
MRN	Mobile Relay Network
NAT	Network Address Translation
NDP	Neighbor Discovery Protocol
NMT	Nordic Mobile Telephone
NPSTC	National Public Safety Telecommunications Council
NTT	Nippon Telegraph Telephone
OFDMA	Orthogonal Frequency Division Multiple Access
OL/SDN	Overlay SDN
ONF	OpenFlow Networking Foundation
PCRF	Policy Charging Rule and Functions
PDA	Personal Digital Assistant
PMI	Precoding Matrix Indicator
RDNSS	Recursive DNS Server
RN	Relay Node
RRM	Radio Resource Management
RUE	Relay User Equipment
SAEGW	System Architecture Evolution Gateway
SC-FDMA	Single Carrier-Frequency Division Multiple Access
SCS	Slice Creation Service
SDN	Software-Defined Networking
SDR	Software-Defined Radio
SGSN	Servie GPRS Support Node
SR	Scheduling Request
TCP	Transmission Control Protocol
TIB	Tolerable Interference Broadcasting
TSG	Technical Specification Group
UDP	User Datagram Protocol
UMTS	Universal Mobile Telecommunication System
V2I	Vehicle-to-Infrastructure
V2V	Vehicle-to-Vehicle
VPL	Vehicle Penetration Loss
WLAN	Wireless Local Area Network
WRAN	Wireless Regional Area Network

Chapter 1
Introduction

Exchanging information electronically has almost become a necessity in today's world. Starting with the traditional telephone networks, electronic communication has evolved significantly over the years. Today we have networking technologies that cater for information exchange at all geographical scales. We have satellite communication and cellular networks that provide long range communication services as well as wireless local and personal area networks for data exchange over a shorter distance. Body Area Networks have recently evolved that provide data exchange over an even smaller expanse. The progress in telecommunications has been rapid, and the interest in the same continues to grow even today. This book tries to capture some of the most recent developments in communication networks.

The communication networks have evolved in two main directions. Data communication over the network dates back to 1970s when Advanced Research Projects Agency (ARPA) introduced ARPANET. In the beginning, ARPANET connected four academic units namely University of California at Los Angeles, University of California at Santa Barbara, Stanford Research Institute and University of Utah. The initial target was to share research results among these academic institutions (Forouzan 2007). The ARPANET initiative has now turned into the Internet, which connects millions of devices worldwide. Soon after the introduction of Internet, Cerf and Kahn (1974) developed protocols and architecture for its large scale deployment. While Internet was only beginning to grow, the cellular networks evolved from the other end as wireless systems for serving the mobile devices. The cellular networks were meant to carry voice when they were first introduced in early 1980s. However, the present days cellular networks support voice as well as data. This book focuses on describing some of the most innovative progress made in cellular and IP networks by citing relevant recent works.

S. F. Hasan, *Emerging Trends in Communication Networks*,
SpringerBriefs in Electrical and Computer Engineering,
DOI: 10.1007/978-3-319-07389-7_1, © The Author(s) 2014

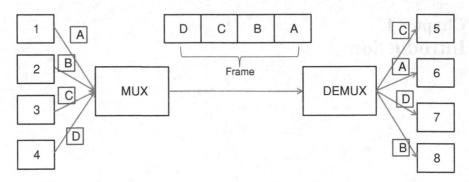

Fig. 1.1 MUX and De-MUX lets multiple users transmit data on the same dedicated link

1.1 Preamble

Networking enables communication over a long distance. The most fundamental technique that allows sending and receiving data over a long distance is switching. Switching is concerned with transmitting data from source to destination over multiple hops. Starting with the legacy circuit switching, considerable progress has been made to improve the switching mechanism. The telephone networks are still based on circuit switching principles, which previously established one dedicated link between a source and destination pair. Since one link housed only one communication session, several links were required to accommodate a large number of users. Time-Division Multiplexing (TDM) based circuit switching solved this problem. The so-called MUX was used to combine data from multiple sources and transmitted the resulting *frame* on a single link. A De-MUX was used at the other end to separate the data contained in a frame and to direct it to their respective destination (see Fig. 1.1). This traditional approach catered for users that had similar data rate requirements, because the transmission opportunities for all users were fixed. Multirate circuit switching allows the allocation of different transmission opportunities to different sources. This idea enables users that require different data rates to coexist on the same link.

While circuit switching has proven to be very successful in traditional voice-based communication systems, it cannot cater for bursty sources. A source is characterized as bursty if its ratio of peak data rate to average data rate is high. Such sources may transmit data only during a small amount of time, while remaining silent during the rest. Since circuit switching allocates fixed transmission opportunities, a bursty source uses its opportunity for a small amount of time only. During the rest of the time period, the transmission opportunity is wasted. This phenomenon is often referred to as the under utilization of the network resources. It can be shown easily using any packet sniffing software that the traffic carried over the Internet is bursty in nature.

Circuit switching has proven to be an inefficient choice to support bursty. Therefore, another switching mechanism had to be introduced.

Packet switching has grown immensely popular in supporting data oriented services because it can easily accommodate bursty users. Unlike circuit switching, packet switching does not allocate fixed transmission opportunities to the users. Instead, all users are free to transmit their data whenever required. Several multiple access techniques have been proposed that resolve the contention between users that try to access the network at the same time. Nevertheless, since users are not allocated transmission opportunities, the resources do not get wasted even if a user is not transmitting data. On the other hand, circuit switching limits the number of users accessing the network at a time. Hence, provision of Quality of Service (QoS) is easier. While packet switching efficiently accommodates bursty users, it cannot guarantee a certain level of QoS. QoS provision becomes a challenge in packet switched networks because there exists no control on the number users or on the amount of data they transmit. Certain packet switched networks use the admission control mechanisms to provide a certain level of QoS. However, this is not always the case. Despite its apparent disadvantages in terms of providing QoS guarantees, the current Internet infrastructure is built around the principles of packet switching. On the other hand, the present day cellular networks use the ideas adopted from circuit switching. This book examines this fact later in Chap. 5 , where the frame structure of LTE networks is discussed in detail.

Both cellular and packet switched IP networks have made considerable progress over the years. The most recent trends in both areas have been covered in this book.

1.2 Book Structure

1.2.1 Scope and Outline

The topics covered in this book can be classified into two sections. The first section deals with the developments in the IP network while the other section highlights the progress being made in the cellular networks.

The first section comprises of Chaps. 2–4. Chapter 2 gives an overview of version 6 of the IP network. The modifications related to the IP addresses and packet headers proposed by IPv6 have been covered in this chapter. Chapter 3 covers the detail of defining an IP network with the help of software. The so-called Software-Defined Networking (SDN) has been shown to increase the flexibility of networks by making them programmable. Chapter 4 of this book discusses the opportunistic access of the communication networks. It presents three innovative communication scenarios that are opportunistic in nature. These topics include Cognitive Radio, Vehicular Communication and Mobile Relay Networks. The cellular networks are discussed in Chaps. 5 and 6. Chapter 5 highlights the LTE network and its associated issues. This chapter also highlights the advances proposed in the so-called LTE-Advanced network. The concept of Fifth Generation (5G) cellular networks has been described in Chap. 6. This chapter starts by establishing the motivation of using 5G networks. It describes UE Relaying and Device-to-Device communication as means to implement a 5G cellular system. This book has been summarized in Chap. 7. Each chapter in this book ends with a chapter summary.

1.2.2 Intended Audience

This book is meant to provide an insight into the state of the art of the recent research and development trends in communication networks. It benefits anybody who is interested in getting familiarized with the upcoming innovations in telecommunications. It is best suited for senior undergraduate students, graduate students and early stage researchers. This book can assist the students in selecting an area of future research activity. It may also be helpful to students who wish to expand their knowledge domain beyond their own specific areas of interest. This book assumes that the reader has elementary knowledge of communication networks. While it still explores advanced topics in networking, it does so in relatively simpler terms. Most of the description is supplemented with illustrative diagrams.

References

V.G. Cerf, R.E. Kahn, A protocol for packet network intercommunication. IEEE Trans. Commun. **22**(5), 637–648 (1974)

B.A. Forouzan, *Data communications and networking* (McGraw Hill, New York, 2007)

Chapter 2
IPv6 Networks

The mechanisms with which we exchange information have evolved significantly over the years. Despite its delays, the postal mail used to be the only method of communication a long time ago. Public switched telephony later made the process considerably faster and less laborious. More recently, the mobile cellular networks and IP networks allow information exchange with a much higher speed and reliability. All mechanisms that allow sending and receiving information require a supporting infrastructure that handles data delivery. For example, postal mails are sent from one geographical location to another through a series of exchange offices. The circuit-switched telephone networks also have dedicated network infrastructure. Similarly, an Internet Protocol (IP) network serves as infrastructure for most packet switched communication networks. Typically, an IP-based infrastructure comprises of hosts that send and receive information through a series of interconnected routers and gateways.

This chapter focuses on discussing the IP network in detail. This discussion is important because of two main reasons. Firstly, the idea of All-IP-Networks (AIN) is rapidly gaining popularity (Shin et al. 2013), which means that IP network shall operate as the backbone for all kinds of communication networks. Even the most recent technologies, like LTE, also provides means to connect to the IP network. Secondly, we are gradually shifting towards a new version of IP, namely IP version 6, which proposes several modifications in the legacy IPv4. Examining these modifications is important for understanding the operation of future IP networks. This chapter starts with a discussion on legacy IP-based networking infrastructure. After building the necessary foundation, this chapter discusses IPv6 in detail. The changes proposed by IPv6 in addressing and routing mechanisms have been examined in depth. The next two chapters of this book discuss innovative ways in which an IP network can be used and further enhanced. Chapter 2 examines the advantages of modifying the IP network by making it programmable while Chap. 3 explores the opportunistic use of the IP network.

S. F. Hasan, *Emerging Trends in Communication Networks*,
SpringerBriefs in Electrical and Computer Engineering,
DOI: 10.1007/978-3-319-07389-7_2, © The Author(s) 2014

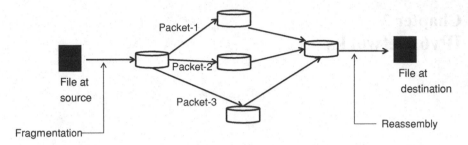

Fig. 2.1 A file is broken down into packets at the source (fragmentation) and put back together at the destination (reassembly). Each packet can traverse a different path

2.1 Basics of IP Networks

An IP network is a packet switched network that breaks the data to be sent into small packets. These packets are sent across a series of forwarding devices called routers. Routers receive packets at one interface, determine the next router using the so-called routing algorithms and send them out on an outgoing interface. Figure 2.1 shows the file transfer in a typical packet switched network. Note that packets originating from a file traverse through multiple routers before getting received at the destination via multiple paths. This method of data delivery is referred to as the datagram approach. In the datagram approach, a file to be sent is *fragmented* at the source into smaller packets. These packets are received at the destination and then *reassembled* back together in order. Fragmentation and reassembly have also been shown in Fig. 2.1.

Unlike circuit switching, packet switched IP networks do not reserve a dedicated path between source and destination. This feature of packet switching makes it suitable for supporting bursty traffic. A typical bursty traffic source does not transmit information consistently. Instead, there are some periods during which the amount of transmitted data is high whereas in other periods no (or very little) data is transmitted. Since most of the internet traffic is bursty in nature, packet-switched IP networks have become extremely popular. A simple packet capture on a live network can verify the bursty nature of internet traffic. Figure 2.2 shows internet traffic captured during a brief web browsing session. It is obvious that data transfer over the network is not consistent. If a dedicated path is reserved for a bursty source, as is the case in circuit switching, it shall remain unused during the periods when no data is transmitted. This results in under utilization of the network resources. Packet-switched networks address this issue by fragmenting a file into packets and deliver them to the destination over multiple hops.

Since multiple devices are involved in the data transfer process, it becomes necessary to uniquely identify the devices that handle data delivery. The identifiers used to name the devices in an IP network are referred to as IP addresses. The current infrastructure supports version 4 of IP (IPv4), in which an IP address is 4 bytes (32 bits) long. The hierarchy of an IP address is such that the first few bits identify the first hop router while the remaining bits of the address identify the device itself.

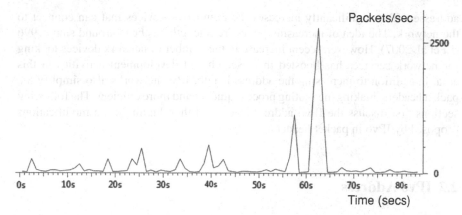

Fig. 2.2 Live packet capture shows the bursty nature of internet traffic

Fig. 2.3 Host and network
parts in an IP address

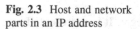

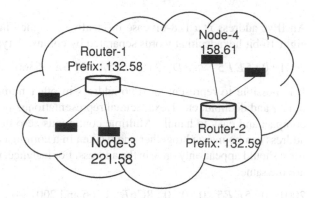

Hence an IP address has a network part and a host part. The IP addresses are hier-archically similar to the telephone numbers which use the first few digits to identify the area and the rest of the digits to identify the phone itself. Figure 2.3 shows host and network parts of IP addresses such that the first 16 bits represent the network. The IP address of Node-3 in Fig. 2.3 is the combination of network and host parts, i.e. 132.58.221.58. All other devices connected to Router-1 have the same host part which is often referred to as the network prefix. The number of bits assigned to repre-sent network prefix is known as the prefix length. In Fig. 2.3, all devices have prefix length of 16 bits. A node connected to a certain router uses its network prefix as the host part of the IP address. Node-4 in Fig. 2.3 uses the network prefix of Router-2, hence its IP address is 158.61.132.59.

The address space of an IPv4 network can support 2^{32} devices without address duplication. A few techniques, such as Network Address Translation (NAT), may be employed to accommodate more devices on the network (Wing 2010). However, the number of devices looking for a network connection has risen so much recently that a more aggressive addressing approach is required. By supporting 128-bit IP addresses, IPv6 significantly increases the number of devices that can connect to the network. The idea of increasing the address length has been around since 1996 (Li et al. 2007). However, recent increase in the number of network devices looking for network services has boosted the research and development activities in this area. In addition to increasing the address length, IPv6 networks also simplify the packet headers, making the routing process quicker and more efficient. The following sections first discuss the IPv6 address space and then highlight the modifications proposed by IPv6 in packet headers.

2.2 IPv6 Address

2.2.1 Format

An IPv6 address is a 128-bit case insensitive unsigned integer. It is represented as eight 16-bit hexadecimal words separated by colons. A typical example looks like:

$2001 : 0 : 5EF5 : 79FD : 24C8 : 85A : 8C6E : 6366.$

It is possible to represent an IPv6 address in other number systems, for example binary and decimal, etc. Hexadecimal representation is preferred because it is more compact and easy to handle. Multiple contiguous zero fields appearing in the IPv6 address can be merged together and written in a compressed form. The compressed form should appear only once in the address. For instance, the following IP addresses are the same:

$2001 : 0 : 5EF5 : 0 : 0 : 0 : 8C6E : 6366$ and $2001 : 0 : 5EF5 :: 8C6E : 6366$

where :: represent contiguous zero fields.

An unspecified address is an IPv6 address with all fields put to zero. This address is useful in the autoconfiguration process discussed later. Sample examples of an unspecified address are:

$0 : 0 : 0 : 0 : 0 : 0$ or equivalently ::

The complete representation of IPv6 address often requires its prefix length. An IPv6 address with prefix length of 64 bits is represented as:

$2001 : 0 : 5EF5 : 79FD : 24C8 : 85A : 8C6E : 6366/64.$

Such a representation indicates that the first 64 bits of the address identify the network while the rest identify the host. In IPv6 terminology, the network part of the address is referred to as the subnet prefix whereas the host part is known as the interface ID.

The interface ID is related to the device's MAC (or physical) address. An interface ID can be configured either manually or automatically. According to IEEE EUI-64, the interface ID is 64 bits long (Li et al. 2007) as shown in Fig. 2.4. Automatic

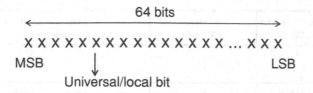

Fig. 2.4 Interface ID of IPv6 address according to IEEE EUI-64 format

configuration of interface ID requires (i) adding *FFEE* (in hexadecimal) in between the device's MAC address and (ii) inverting the universal/local bit. This converts 48 bit MAC address into 64 bit interface ID. The universal/local bit identifies whether the interface ID is defined locally or globally. The universal/local bit is set if an IPv6 address is locally defined.

The present day network devices often come with more than one network interfaces. For instance, a smart phone will have separate interfaces for 3G, Wi-Fi and WiMAX, etc. IPv6 can identify the device having multiple interfaces as well as the individual interfaces within the device. It uses Device User ID (DUID) to name a device that may have multiple interface cards. It also allows the use of Interface Association ID (IAID) in order to name each interface within the device.

2.2.2 Types of Addresses and IPv6 Messages

An IPv6 address is classified into three main types: Unicast, Multicast and Anycast address. A unicast IPv6 address identifies a single network interface to which packets are delivered. A multicast address, on the other hand, identifies a group of interfaces. The packets destined to a multicast address are received by all interfaces present in the group. It is also possible to send packets to a group such that they are received only by a single node. The receiving node may be selected by a routing algorithm based on how close it is to the source. This type of addressing is known as anycast addressing. Anycasting has two usage restrictions: (i) an anycast address must not be used as the source address, and (ii) anycast addresses are assigned to the routers.

IPv6 addresses can also be categorized as global and link-local addresses. Both kinds of addresses have different formats, as shown in Fig. 2.5 (Li et al. 2007). As the name indicates, a node uses global address to communicate over the network. This kind of an address is unique over the entire network. Link-local address in IPv6 is recognized by the prefix FE80 in hexadecimal. The link-local address is used in the Neighbor Discovery Protocols (NDP) for a variety of purposes (Alsa'deh and Meinel 2012). The Neighbor Discovery protocol helps nodes in detecting the available points of attachments. The protocol has five main purposes: router solicitation, router advertisement, redirection, neighbor solicitation and neighbor advertisement. All five NDP messages are carried in the Internet Control Messaging Protocol (ICMPv6) packets.

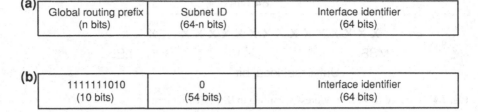

Fig. 2.5 Local and global IPv6 address

Router Solicitation messages are used by the nodes to query the information about the available routers. They are commonly used to identify a router when the device connects to the network. Routers often indicate their presence via the router advertisement messages. These messages are sent periodically and they contain the network prefix that is currently used by the router. The Router Advertisements used by NDP carry the link-layer address thus removing the need for resolving router's link-layer address. Router advertisements also carry the prefixes for every link, and allow address autoconfiguration. A link in IPv6 network can be associated with multiple prefixes. The information on all prefixes in use by the link are broadcast in the router advertisements (Narten et al. 2007).

The neighbor solicitation messages are used to query a neighboring node. It can be seen as a replacement of Address Resolution Protocol (ARP) request that verifies whether the neighboring node is active. Neighbor advertisement message is sent in response to the neighbor solicitation messages. The redirect messages are used by the routers to inform network devices about a possible alternate path to their respective destinations. These are particularly useful when a path faces failure or congestion.

2.2.3 Stateless Autoconfiguration

In IPv4 networks, devices dynamically acquire the IP address using the Dynamic Host Configuration Protocol (DHCP) (Blank 2002). This kind of address allocation is known as stateful autoconfiguration. IPv6 networks, on the other hand, allow the use of stateless autoconfiguration. In stateless autoconfiguration, DHCP server is not used and the address is generated by the node automatically.

The stateless autoconfiguration process proceeds as follows. A node first generates a link-local address as discussed in the previous section. This address is not assigned to an interface until its uniqueness is verified. In order to verify whether the link-local address is unique, Duplicate Address Detection (DAD) tests are performed. Several methods are available for performing DAD test (Thomson and Narten 2007). One simple method is that the node sends Neighbor Solicitation message with the address to be tested as the target address. If the address is already in use, the node already using the address shall send a response. If an address is not found to be unique, autoconfiguration shall not proceed further and the address shall have to be assigned using DHCPv6 server (or manually). On the other hand, if the link-local address is found to be unique, it is assigned to an interface. In the next step, the node listens to the Router Advertisement messages from the neighboring routers. As mentioned earlier, these messages will contain the network prefixes used by the advertising router. Using the information contained within the advertisement messages, the node generates an IPv6 global address. The uniqueness of this address may also be checked using DAD mechanisms. Once a node gets an IPv6 address, it can initiate DNS configuration. While IPv4 networks use dedicated DNS servers, DNS configuration is a little different in IPv6.

2.2.4 DNS Configuration

Domain Name System (DNS) servers are used in traditional IP networks to convert domain names into IP addresses. The information about the available DNS servers and their search list is contained in Router Advertisement and DHCP messages (Park et al. 2013). The recursive DNS server (RDNSS) and DNS search list (DNSSL) options have recently been introduced to carry DNS information. The RDNSS option contains the IPv6 address(es) of the recursive DNS servers that may be contacted for name resolution. On the other hand, DNSSL contains the domain names of DNS suffixes. In cases when RDNSS and DNSSL options are available from both sources Router Advertisement messages and DHCP, it is recommended to store at least three RDNSS addresses or DNSSL domain names (Jeong et al. 2010). Another possibility is to store the well known DNS server addresses in the IPv6 hosts's registry. This method is useful because, unlike other approaches, DNS configuration shall not require traffic exchange over the network. However, since the Internet Assigned Numbers Authority (IANA) have not assigned well known addresses to the DNS servers, this method cannot be used just yet. Nevertheless, (Park et al. 2013) consider this approach as a suitable candidate for DNS configuration in addition to other approaches.

This concludes our discussion on IPv6 addresses and their use in different operations. Most of the definitions covered so far can be seen in the IP configuration tool of the Disk Operating System (DOS). Figure 2.6 shows IP related information of a device that is connected to a network. The device is currently using IPv4 address because we have not switched to a standalone IPv6 network yet.

2.3 IPv6 Packet Headers

A packet in an IP network comprises of a header and a payload. The payload contains the actual data to be sent while the header contains information required for data delivery. In addition to increasing the address space, IPv6 also simplifies the packet

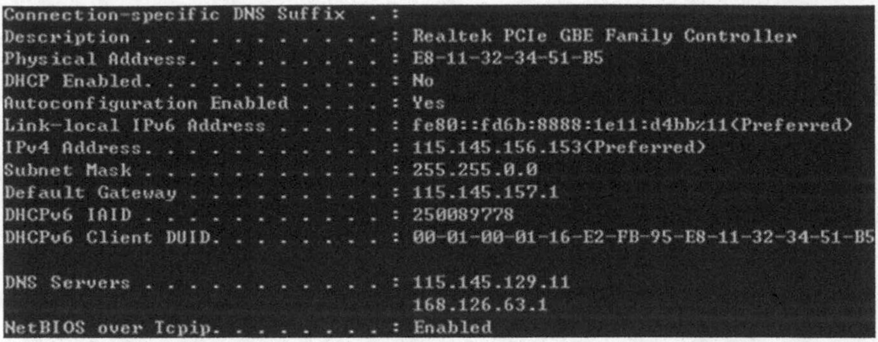

```
Connection-specific DNS Suffix  . :
Description . . . . . . . . . . . : Realtek PCIe GBE Family Controller
Physical Address. . . . . . . . . : E8-11-32-34-51-B5
DHCP Enabled. . . . . . . . . . . : No
Autoconfiguration Enabled . . . . : Yes
Link-local IPv6 Address . . . . . : fe80::fd6b:8888:1e11:d4bb%11(Preferred)
IPv4 Address. . . . . . . . . . . : 115.145.156.153(Preferred)
Subnet Mask . . . . . . . . . . . : 255.255.0.0
Default Gateway . . . . . . . . . : 115.145.157.1
DHCPv6 IAID . . . . . . . . . . . : 250089778
DHCPv6 Client DUID. . . . . . . . : 00-01-00-01-16-E2-FB-95-E8-11-32-34-51-B5

DNS Servers . . . . . . . . . . . : 115.145.129.11
                                    168.126.63.1
NetBIOS over Tcpip. . . . . . . . : Enabled
```

Fig. 2.6 IP configuration tool showing IPv4 and IPv6 information

header. The information contained in the packet header is processed by routers to determine where a certain packet is to be sent. A large and complicated packet header would naturally require larger processing time. Various simplifications in packet headers have been proposed by IPv6 that increase the routing speed and allow quicker delivery of information. One of the major modifications in the packet header is the introduction of extension headers that are appended next to the base header. The base header only carries the necessary routing information while all supplementary information is available in the extension headers. A base header may have multiple extension headers. It is not necessary to use extension headers at all times. They are used whenever needed. The following discussion covers the base header and extension headers one by one.

2.3.1 Base IPv6 Header

The base IPv6 header is 40 bytes in length and contains fewer fields in comparison with the legacy IPv4 header (Brown and Parenti 2002). IPv6 header is the simplified version of IPv4 in that only necessary fields have been retained. Figure 2.7 shows the new IPv6 header with 128-bit source and destination addresses. As the name indicates, the version field in the base header specifies the version of IP (version 6 in this case) while the payload length indicates the size of data contained in the packet. The next header field specifies which extension header(s) follow the base header. If there are no extension headers, the next header field in the base header specifies the transport layer protocol that carries the IP packet. Common transport layer protocols include TCP and UDP etc. The priority field specifies the traffic class. This feature is used when the network faces congestion. If the congestion is severe, network has to drop a few packets to get things back to normal. Based on the priority level specified in the IPv6 header, network decides whether a particular header must be dropped during congestion. The packets containing real-time traffic generally have a higher

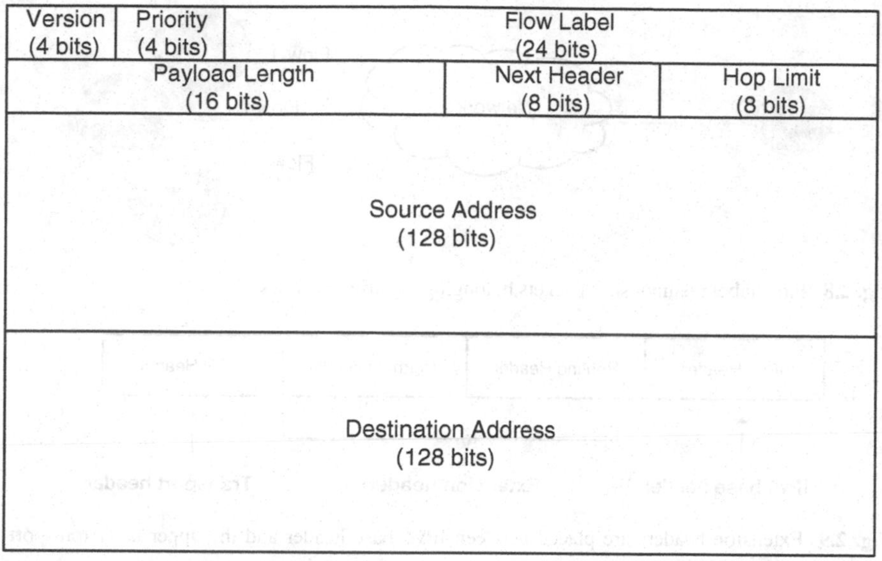

Version (4 bits)	Priority (4 bits)	Flow Label (24 bits)		
Payload Length (16 bits)			Next Header (8 bits)	Hop Limit (8 bits)
Source Address (128 bits)				
Destination Address (128 bits)				

Fig. 2.7 IPv6 Header

priority and are seldom dropped during congestion. It is pertinent to mention that 3rd Generation Partnership Project (3GPP) has classified all applications into four categories: Background, Interactive, Streaming and Conversational. Each of these applications require a different quality of service and priority from the network. These priority requirements may be communicated to the network using the priority field. The hop limit field in the base header is used to identify and drop the packets that are stuck in indefinite loops. Upon receiving the packet, every intermediate router decreases the value specified in the hop limit by 1. A packet is dropped if its hop limit becomes 0 before it reaches its destination.

The flow label field in the IPv6 header specifies the packets belonging to the same *flow*. All packets in the same flow are treated in a similar manner by the intermediate routers. A router processes one packet from a flow and caches the results. These results are used to process the rest of the packets belonging to the same flow. This reduces the computational burden on the intermediate routers and speeds up the routing process. A flow may be defined in a variety of ways. For example, a flow may simply comprise of all packets originating from a certain source and destined to a certain destination as shown in Fig. 2.8.

2.3.2 Extension Headers

In legacy IPv4 networks, the IP header is followed by the transport layer header (TCP, UDP, etc). On the other hand, in IPv6 networks, there may be optional extension

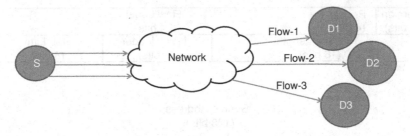

Fig. 2.8 Flow label distinguishes packets belonging to different flows

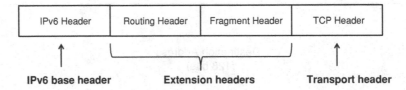

Fig. 2.9 Extension headers are placed between IPv6 base header and the upper layer transport header

Table 2.1 Extension headers specified for use in IPv6 (Loshin 2004)

Extension headers	Next header value
Hop-by-hop header	0
Routing header	43
Fragment header	44
Encapsulating security payload	50
Authentication header	51
No next header	59
Destination options header	60

header(s) between the IPv6 header and the transport layer header. If multiple extension headers are available, they are processed in the same order as specified by the source. Figure 2.9 shows a typical case where an IPv6 header is followed by two extension headers, which are then followed by the TCP header. Six extension headers have been specified for use in IPv6 networks. These have been tabulated in Table 2.1 (Loshin 2004). The value of the next header field to be used in the preceding header for each has also been given. Note that the table also contains an extension header called *No Next Header*. This is an imaginary header that implies that no extension headers are present. The use of such a header is also optional.

The hop-by-hop header appears immediately after IPv6 base header and is processed by all intermediate routers as well as the source and destination. On the other hand, the destination options header is processed only at the destination. The routing header is used at the source, which specifies the routers that are to be encountered by the packet. The number of intermediate routers is given in the Segment Left field of the header. The value contained in this field is decremented by 1 by every

Table 2.2 The fields used in routing header and their description

Fields	Description
Next header	Identifies the protocol header that follows
Header length	Specifies the length of the routing header
Routing type	Several types of routing headers have been specified
Segments left	Identifies the no. of route segments to be visited
Type specific data	Contains data according to routing type in use

intermediate router. As the name indicates, Fragmentation header is used for fragmenting and reassembling the packet. While a dedicated header for fragmentation is available, it must be noted that fragmentation is discouraged in IPv6 specifications. The Destination Options header contains variable length fields to enforce a few actions on the packet as it reaches the final destination. The Encapsulating Security payload and Authentication headers are meant to ensure security and privacy. These issues are out of this book's scope. A detailed discussion on IPv6 headers can be found in (Loshin 2004).

It is obvious from this discussion that different devices within the network are responsible for processing selected extension headers only. In IPv6, all devices do not have to process everything that comes their way. This makes processing headers less complicated than IPv4. In the following, we briefly discuss the routing header as an example of the extension headers. The purpose of this discussion is to show how extension headers are used in routing packets over the network.

The routing header contains the list of intermediate routers that a packet shall encounter as it moves from its source to the destination. These intermediate nodes are referred to as the route segments. The fields included in the routing header and their descriptions have been shown in Table 2.2. In addition to these fields, the routing header also contains source and destination addresses and the reserved field. The purpose of each of these fields has been explained in Fig. 2.10 (Li et al. 2007). The following explanation is for Type 0 routing headers. The figure shows that there are two route segments between source and destination. The source and destination addresses are changed every time the header reaches a route segment. As the packet reaches its first route segment, the segment left field is decremented by 1, and source and destination addresses are changed according to the next hop route segment. Note that the routing header contains the complete list of route segment addresses until it reaches its destination. At the destination, the segment field becomes 0 indicating that no more routing is required. The destination recognizes its address in the destination address field of the packet and accepts it as received.

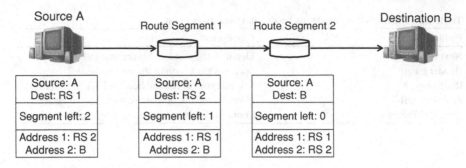

Fig. 2.10   An example to show the use of routing header

2.4 Migration to IPv6

Today's internet is built on IPv4 backbone. When the idea of IPv6 was introduced back in 1996, migration to IPv6 was considered only as a suggestion. Therefore, the process of migrating to an IPv6-only network has been very slow. There are three main players migrating from IPv4 to IPv6. The first is the service providers, who will migrate only when their service capacity is affected by the lack of IP addresses. The second are the enterprises, who will migrate only when their reachability and presence on the internet gets affected because of IPv4. Finally, there are end users, who simply do not care which address they are using (Kaur 2013). Both service providers and enterprises are realizing the need for migrating to IPv6. IPv6 was globally launched on 6 June 2012 with the help various participants including vendors, website and network operators (IPv6Launch 2012). The transition from IPv4 to IPv6 has begun and, more notably, has become almost mandatory due to the lack of IPv4 addresses.

Despite having sparse deployment of IPv6 around the world, much of the present IP infrastructure still supports IPv4. The complete migration to IPv6 will of course take time. The share of IPv4 in the current networking infrastructure is much larger, which is expected to decrease with the increasing deployment of IPv6 networks. Until this migration process is complete, IPv6 will have to co-exist with IPv4 infrastructure. There are three main strategies that have been adopted to allow this co-existence, as briefly highlighted in the following:

- Dual Stack: This approach allows a device to have two stacks, one each of IPv4 and IPv6 protocols. Dual stack is the easiest to implement and various operating systems have this feature built in already.
- Header Translation: This method uses algorithms to translate one header format into another. This is the least popular method which may also give rise to several complications.
- Tunneling: Tunneling is the most popular approach which allows IPv6 traffic to be encapsulated in IPv4 packets (and vice versa) when the traffic is passing through the IPv4 network. The intermediate routers shall treat all packets as v4. The IPv4 headers will be removed once the packets enter IPv6 domain.

2.5 Summary

This chapter describes the next generation of IP networks. IPv6 makes two main modifications in the legacy IPv4 protocol. Firstly, it increases the IP address length from 32 to 128 bits. This obviously increases the address space and hence the number of devices that can simultaneously access the network without address duplication. Secondly, IPv6 uses extension headers to simplify the base IP header. All extension headers used in packet delivery are not processed by all devices on the network. Respective devices process their concerned headers only, which speeds up the routing process. The current network, despite having larger portion of IPv4, is gradually migrating towards IPv6. In a few years time this migration shall become necessary, specially when techniques like machine-to-machine communications are implemented.

References

A. Alsa'deh, C. Meinel, Secure neighbor discovery: review, challenges, perspectives, and recommendations. IEEE Secur. Priv. **10**(4), 26–34 (2012)

A.G. Blank, TCP/IP Jumpstart: Internet protocol basics. (John Wiley and Sons, 2002)

E. Blanton, S. Chatterjee, S. Gangam, S. Kala, D. Sharma, S. Fahmy, P. Sharma. Design and implementation of the S3 monitor network measurement service on GENI. in 4th International Conference on Communication Systems and Networks (2012)

S. Brown, E. Parenti. Configuring IPv6 for cisco IOS. Syngress (2002)

Cisco. Software-defined networking: why we like it and how we are building on it. White Paper. (2013)

M. P. Fernandez . Comparing openflow controller paradigms scalability: Reactive and proactive. in International Conference on Advanced Information Networking and Applications (2013)

FloodLight. Project Floodlight: Open source software for building software-defined networks (2014) Available online at http://www.projectfloodlight.org/

GENI. Geni experiment and assignment repository (2013a) http://groups.geni.net/geni/wiki/GENIExperimenter/ExampleExperiments

GENI. Global environment for networking investigation. (2013b) Available online at geni.net.

S. Hares, R. White, Software-defined networks and the interface to the routing system (I2RS). Unknown Journal **17**(4), (2013)

Heller, Sherwood, McKeown. The controller placement problem. in ACM Hot SDN (2012)

IPv6Launch. (2012) http://www.worldipv6launch.org/. (available online)

M. Jarschel, F. Lehrieder, Z. Magyari, R. Pries.. A flexible openflow-controller benchmark. european workshop on software defined networking. (2012)

J. Jeong, S. Park, L. Beloeil, S. Madanapalli. IPv6 router advertisement options for DNS configuration. IETF RFC 6106 (2010)

G. Kaur. IPv6 Transition and deployment strategies–lessons from the trenches. LightReading webinar (2013)

Kurose, J. F., Ross, K. W., 2008. Computer networking: A top-down approach. IEEE Internet Comput. (Addison Wesley, 2008)

D. Levin, A. Wundsam, B. Heller, N. Handigol, A. Feldmann, *Logically Centralized? State Distribution Trade-offs in Software Defined Networks,*in *Workshop on Hot Topics in SDN, ACM Sigcomm* (2012)

Q. Li, T. Jinmei, K. Shima, *IPv6 Core Protocols Implementation* (Elsevier, Morgan Kauffman Series of Networking, 2007)

P. Loshin. IPv6: theory, protocol and practice. Morgan Kaufmann (2004)

N. McKeown, T. Anderson, H. Balakrishnan, G. Parulkar, L. Peterson, J. Rexford, S. Shenker, J. Turner. Openflow: enabling innovation in campus network. in ACM SigComm (2008)

D. Meyer, The software-defined-networking research group. IEEE Internet Comput. **17**(6), 84–87 (2013)

Moy, J. T., OSPF: Anatomy of an Internet Routing Protocol (Addison-Wesley Professional, 1998)

T. Narten, E. Nordmark, W. Simpson, Neighbor discovery for IP version 6 (IPv6). IETF RFC 4861 (2007)

S. Ortiz, Software— defined networking: on the verge of a breakthrough? IEEE Comput. **46**(7), 10–12 (2013)

S. Park, J. Jeong, C.S. Hong, DNS configuration in IPv6: approaches, analysis, and deployment scenarios. IEEE Internet Comput. **17**(4), (2013)

T. Sedmak, Internet2 and networking industry partners drive innovation and development of SDN and openflow applications for 100G network. available online at internet2.edu (2013)

S.A. Shah, J. Faiz, M. Farooq, A. Shafi, S.A. Mehdi, An architectural evaluation of SDN controllers. IEEE International Communications Conference's Next Generation Networking Symposium **10**, 3504–3508 (2013)

D.H. Shin, D. Moses, M. Venkatachalam, S. Bagchi, Distributed mobility management for efficient video delivery over all-IP mobile networks: competing approaches. IEEE Network **27**(2), 28–33 (2013)

V. Thomas, GENI: Exploring networks of future.in *15th GENI Engineering Conference* (2012)

S. Thomson, T. Narten. IPv6 Stateless address autoconfiguration. IETF RFC 4862 (2007)

TraceRoute. Available online at. Network-Tools.com. (2013)

S.J. Vaughan-Nichols, Openflow: The next generation of the network. IEEE Computer **44**(8), 13–15 (2011)

D. Wing, Network address translation: extending the internet address space. IEEE Internet Comput. **14**(4), 66–70 (2010)

X. Yin, Huang, S. Wang, D. Wu, Y. Gao, X. Niu, M. Ren, H. Ma . Software defined virtualization platform based on double-flowVisors in multiple domain networks.in 8th International Conference on Communications and Networking in China (2013)

Chapter 3
Software-Defined Networking

3.1 Introduction

In a typical packet-switched network, finding the path of a packet that is moving from one node to another is known as routing. It is easy to imagine that routing is at the heart of IP-based communication networks. Without effective routing mechanisms, it is impossible to exchange information between distant nodes. Years of research in routing has resulted in the development of several routing algorithms. In the conventional paradigm, these algorithms are executed on hardware devices known as routers. Using a certain routing algorithm, each router would decide where to direct the incoming packets. The routing mechanism in the current IP network is distributed, such that each router makes its own routing decisions. Efforts are underway to centralize the prevailing routing mechanism. This has been made possible by installing a central device called the controller, which makes the routing decisions for the entire network. By allowing a central controller to manage all routing decisions, all network routers will not have to execute time consuming and computationally expensive routing algorithms. All decisions shall be made in a remote controller and conveyed to the routers. These routers are referred to as the *programmable* routers.

The networks that employ programmable routers to facilitate the routing mechanism are known as Software-Defined Networks. Software-Defined Networking (SDN) became popular when it was experimented and later deployed as part of the Ethane project at Stanford University. SDN got a significant boost in July 2012 when Vmware purchased the SDN vendor Nicira for USD1.05 billion. In terms of development, an open source framework has been initiated by Open Daylight to allow vendors to build various SDN products (Ortiz 2013). More recently, Cisco has also expressed support for SDN (Cisco 2013), which has further fueled the research and development activity in SDN. This chapter is dedicated to a detailed discussion on software-defined networking, its basic principles and a few research challenges.

This chapter starts by looking at the limitations of the current routing methods. It then explores the changes suggested by SDN in the current paradigm and their

S. F. Hasan, *Emerging Trends in Communication Networks*,
SpringerBriefs in Electrical and Computer Engineering,
DOI: 10.1007/978-3-319-07389-7_3, © The Author(s) 2014

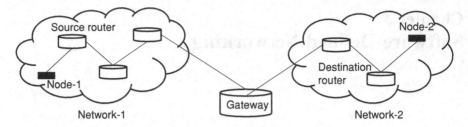

Fig. 3.1 Source, destination and intermediate routers serving information exchange between Node-1 and Node-2

advantages. After discussing the preliminaries, this chapter describes SDN in more technical detail. While the main discussion remains at the Network layer, a few remarks on Software-defined physical layer have also been included towards the end of this chapter.

3.2 Conventional Routing Paradigm

An IP-based communication network is composed of a series of interconnected routers. Each router serves a set of nodes, which may be desktop computers, laptops, PDAs, etc. The router which is one hop away from an end user is known as the default router for that node. Packets sent by a node are received by its default router and forwarded to the default router of the destination node. The source and destination default routers are connected by a series of intermediate routers. A default router is termed as the source router when it serves the source node. Similarly, it is known as the destination router when it serves the destination node. Figure 3.1 shows the source and destination routers in a network when Node-1 sends packets to Node-2. Routing is therefore defined as the process of forwarding packets between source and destination routers over a typical IP network.

In the current IP network, all routers execute a particular routing algorithm to determine where the incoming packets must be sent. The primary role of a routing algorithm is to determine the next hop router for efficient packet delivery. Programs like traceroute (TraceRoute 2013) can come handy in appreciating the number of routers involved in a certain packet exchange over a network. For example, from author's desktop, the local Yahoo server is about 10 hops away. If a web-fetch request is sent to the local Yahoo server, it will pass through at least 10 routers. This implies that at least 10 routers have to run routing algorithms to serve one web-fetch request made by a user. Considering that there are millions of web users that simultaneously access the network, the current method of routing requires a lot of processing overhead. The conventional method is therefore computationally expensive for routers, and by extension, for the entire networked system.

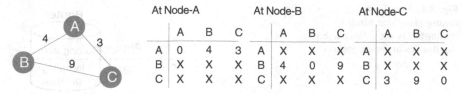

At Node-A				At Node-B				At Node-C		
	A	B	C		A	B	C		A	B	C
A	0	4	3	A	X	X	X	A	X	X	X
B	X	X	X	B	4	0	9	B	X	X	X
C	X	X	X	C	X	X	X	C	3	9	0

Fig. 3.2 A small network with delay values on the branches and routing tables stored at each node

Fig. 3.3 Modified routing table at Node-B

At Node-B

	A	B	C	
A	0	4	3	Must be changed to 7
B	4	0	~~9~~	With next hop A
C	3	9	0	

Let us consider the example of distance vector algorithm to appreciate the amount of work done by the routers in forwarding packets. The purpose of the following discussion is not to explain the algorithm itself. Our main interest is in showing that routers are heavily loaded with a lot of computational burden. A detailed account of DV algorithms can be seen in (Kurose and Ross 2008).

3.2.1 Computational Workload of Routers

Distance Vector (DV) is a distributed, iterative and asynchronous routing algorithm that is commonly employed in communication networks (Moy 1998). The principle focus of DV algorithm is to determine the most effective path between source and destination routers based on cost. Cost may be expressed in many different ways. For example, cost may imply the financial burden associated with transmitting on a link (or a set of links). Cost may also indicate the amount of time delay, link throughput, etc. If cost is determined by time delay, a path between source and destination that yields minimum cost shall be considered most suitable by the DV algorithm.

Consider a small network of three nodes that uses DV algorithm as shown in Fig. 3.2. The figure also shows the cost (in terms of time delay) for all links. Smaller the branch value better the path. When this network is initialized, all nodes determine the time delay between themselves and their neighbors. The measured delay values are stored as tables in each router. For example, the time delay on the link between nodes B and C is 9. Symbol 'X' in Fig. 3.2 implies that the delay values are not known at the time of initialization. After initializing, nodes A, B and C share their tables with each other. Figure 3.3 shows the table of node B, where all entries are now known. Once the entire topology of the network is known, node B realizes that the time delay to node C can be reduced if its packets first arrive at A and then at C.

Fig. 3.4 A router has a
control plane that handles
routing tables and a data plane
that handles incoming packets

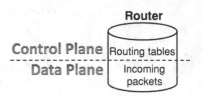

Upon identifying a shorter path between node B and C, node B modifies its table as shown in Fig. 3.3. Nodes A and C also modify their tables accordingly.

This example describes a network that comprises of only 3 nodes. It is easy to imagine the amount of computation required by each router as the number of neighbors increase. A large network may have hundreds of routers. Secondly, the delay values were modified in this example when nodes shared their tables at the time of initialization. In a typical network, delay values may change continuously due to a variety of reasons. For example, if a router stops working, its neighboring routers will have to repeat the entire procedure to determine an alternate route. In wireless networks, link attributes change more often due to time-varying channel conditions. It follows that link values within a network may change continuously and maintaining updated tables put considerable computational burden on the routers. Finally, this much computation only suffices to keep an updated view of the topology. Decoding packet headers, processing them and forwarding data are additional functions of the router.

Based on this discussion, a router can be seen as performing two main tasks. The first is concerned with staying updated about the network topology. This is done by regularly updating the locally stored tables, and is often referred to as control plane operation of routers. The second task is to forward the incoming packets based on these tables. This is often referred to as the data (or forwarding) plane operation of routers. Thus, a router is said to have a control plane and a data plane. The control plane comprises of locally stored tables and the data plane comprises of incoming packets that are forwarded using control plane information. The basic idea of SDN is to decouple the control plane and data plane in order to simplify the overall network operation. A router's control plane and data plane have been shown in Fig. 3.4.

3.3 Decoupling Control and Data Planes

While a conventional router has to manage it's control and data planes both, SDN assigns a router's control plane to an external network device. This external network device is known as the Controller. The controller manages all routing decisions and programs the routing tables contained in the router according to a routing protocol. The operation of a router is remotely managed by a piece of software running in the controller. Figure 3.5 contains a simple graphical representation of a network defined by software. The routers receive packets from the source and send it to the

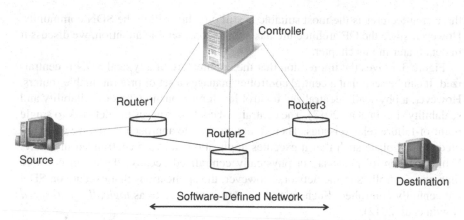

Fig. 3.5 A typical SDN architecture: routers managed by controllers

controller for processing. Controller processes the packet header and updates the tables of the concerned routers accordingly. This remote updating of a router's flow table is referred to as *programming* in the SDN context. The controller also creates a flow within the flow table so that all subsequent packets from the same source get similar treatment from all routers in the network. The controller does not receive all packets from the source. It only receives and processes the first packet belonging to a flow. The rest of the packets are forwarded by the router according to the instructions received from the controller. Note that the routers in SDN are not completely devoid of routing tables. The routers still have flow tables but they no longer have the ability to update them without instructions from the controller. Therefore, a router in SDN is a dumb forwarding device which follows the directions given by the controller. Several architectural designs have been proposed for SDN that are discussed in the following.

3.3.1 Architectural Designs

Meyer (2013) argue that at least three architectural designs are under consideration for use in SDN. The first design approach is to completely separate the control plane from the data plane. This method is adopted in the OpenFlow (OF) design which is discussed in the next section. Another approach, referred to as Control Plane/SDN (CP/SDN) approach, focuses on making the current control plane programmable. While OF requires a new control plane, CP/SDN uses the same control plane but adds the feature of programmability. A new protocol known as Interface to Routing System (I2RS) has been designed (Hares and White 2013) for use in CP/SDN. The third possible architecture is the Overlay SDN (OL/SDN). In OL/SDN, a virtual network is overlaid on the actual network. This virtual overlaid network serves as the control plane for the underlaying data plane. The debate as to which of these three architectures is the most suitable is still ongoing within the SDN community. However, since the OF architecture has gained considerable attention, we discuss it in detail later in this chapter.

Figure 3.5 gives the impression that the architecture of a typical SDN is centralized. It can be seen that a central controller manages a set of programmable routers. However, a physically centralized control has limitations in terms of reliability and scalability (Levin et al. 2012). The centralized control exposes the network to single point-of-failure related issues. Therefore, an alternate approach is to have a distributed control plane such that it executes applications that are centralized in nature. While the control plane is not physically centralized because there can be more than one controllers in the network, however, the applications that execute on SDN are centrally controlled. Such an architecture is referred to as *logically centralized* (Levin et al. 2012).

It must be noted that the degree of separation between control plane and data plane has not been completely agreed upon. In fact, a dedicated research group has been recently established that focuses entirely on the development of SDN. The so-called Software-Defined Networking Research Group (SDNRG) is a part of Internet Research Task Force (IRTF), which examines SDN for all kinds of networks including cellular, home, enterprise, data center and wide-area networks (Meyer 2013). The main idea behind establishing SDNRG is to provide a forum for the researchers where they can discuss, investigate and solve interesting problems related to SDN.

3.3.2 How Is SDN Useful in Research?

The SDN concept is useful in a variety of ways in terms of research. A detailed account of research contributions expected from SDN has been given in (McKeown et al. 2008). One of the main benefits of SDN, which is also highlighted earlier, is that the routers do not have to perform computationally expensive operations. All protocols are executed in the central controller that makes all routing decisions. From the perspective of research and development, SDN is significant because it allows researchers to test their protocols on real production networks (McKeown et al. 2008). Currently, researchers have to rely on simulations or experiments with a limited scope to comment on the suitability of their designs. Some are lucky enough to have access to large-scale networks for experimentation, however, most researchers do not have this luxury. SDN can be helpful in testing new designs as explained in the following.

Suppose a researcher develops a protocol that needs to be tested. Assuming that access to SDN is available, the researcher connects its computer to the SDN-enabled default router. This router (and other SDN-enabled routers) is part of a large network, which is being used by other users as well. The routers in this network have to deal with two kinds of traffic: one is the normal production traffic and the other is the SDN traffic that is being generated by the researcher's computer, as shown in Fig. 3.6. The first packet from the researcher's experimental traffic is sent to the controller.

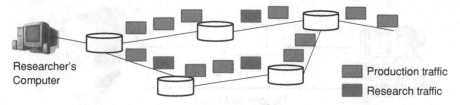

Fig. 3.6 Production and experimental traffic is carried in the same network. Connections between routers and controller are not shown here for simplicity

The controller then executes the researcher's protocol and sets the table entries for all SDN routers within its domain. This way, the packets from the researcher's computer are treated using the protocol design to be tested. The production traffic, on the other hand, is treated as usual through the conventional routing protocols. A suitable metric, such as routing delay etc, can then determine the performance of researcher's design on a fully functional real world network.

Our discussion has so far developed the basic idea of SDN. The following section provides a detailed technical account of SDN using the examples of a well known network that is enabled by software.

3.4 Programmable Networks

A software-defined network is referred to as programmable because the controller can be programmed with different routing protocols. This controller then programs the flow tables of the network routers according to that protocol. In addition to the routing protocol, there are at least two important constituents of a SDN:

- An interconnection of programmable routers, and
- Communication protocol between controller and the programmable routers within its domain.

The first constituent of SDN is the network that allows software-based updating of routing tables. Such a network is an interconnection of routers that can be programmed remotely.

Among other programmable networks, GENI testbed has become increasingly popular as an interconnection of programmable routers that uses SDN. The second important part of SDN is the protocol that allows communication between the controller and the programmable routers. The entire idea of SDN is built around the notion that a controller updates the tables of all routers within the network. This cannot be done without a dedicated light-weight communication protocol that runs between the controller and all programmable routers. While several protocols have been developed for this purpose, OpenFlow protocol has received significant attention.

In the following discussion, we explore the GENI testbed and OpenFlow protocol to strengthen our understanding of SDN. The following discussion is founded on the

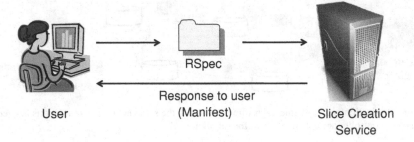

RSpec

Response to user
User   (Manifest)   Slice Creation
                                                                     Service

Fig. 3.7 A user reserves a slice from slice creation service using RSpec

basics that we have already developed and aims to explain SDN in more technical detail.

3.4.1 Global Environment for Networking Investigation

The Global Environment for Networking Investigation (GENI) has been introduced as a virtual laboratory for conducting experiments. In essence, GENI is a big network comprising of hosts, routers and links that are collectively known as resources. Researchers willing to use GENI can reserve a slice of resources to run their experiments (GENI 2013b). A typical slice is a set of GENI resources that may be shared between multiple users. Each user can program his/her reserved slice such that it does not interfere with the slice of the other users (Thomas 2012).

Users can request and create their slices using the Resource specification language (RSpec). The resources requested by the user are defined as an XML-like code. This code is then sent to the Slice Creation Service (SCS). Upon receiving user's request, SCS checks whether the requested resources can be assigned (Blanton et al. 2012). This decision depends on a variety of factors, such as already registered number of slices, number of resources used by each registered slice, GENI thresholds of assigning resources, etc. The response of SCS, also known as the *manifest*, is finally sent back to the user as shown in Fig. 3.7. If the user's request is successful, a set of requested resources is made available to the user. While GENI testbed can be used in a variety applications, a few sample experiments have been given in (GENI 2013a).

A programmable network like GENI certainly requires a set of controllers. A variety of controllers have been made available today for use in a typical SDN. For example, Floodlight has been introduced as a controller for OpenFlow-based SDNs. Its software is freely available online under Apache 2.0 license, which allows third party application development. The code is mainly written in Java and can be easily modified for developing customized applications (FloodLight 2014). Similarly, FlowVisor is a network virtualization software that is also meant for OpenFlow-based SDNs. It acts as a controller by managing the bandwidth, flow tables and utilization of the network routers (Yin et al. 2013). Floodlight is more evolved in comparison with FlowVisor because the later still needs various modifications and features. NOX, Maestro and Trema are among other controllers that have gained considerable popularity over the years.

Several recent works have examined the performance of these controllers in terms of different indicators. It is well known that all controllers have their merits and the choice of most suitable controller depends on requirements. In terms of handling the maximum number of flows, Fernandez (2013) have shown that C++ based NOX controllers outperform the rest for a network with 200 switches. The floodlight controller comes in second with performance very close to the NOX controller. On the other hand, according to the observations recorded in (Shah et al. 2013), the Beacon controller outperforms the rest in terms of supporting a high flow rate. The performance was tested on a 16-node network with different numbers of threads and routers. The observations reported by Fernandez (2013) and Shah et al. (2013) have been made using CBench tool, which evaluates performance at a network level. A new controller evaluation tool has been proposed by Jarschel et al. (2012) that examines each controller on per switch basis. More work is being done on further improving this new benchmark.

3.4.2 How Many Controllers and Where?

With increasing interest in programmable networks, testbeds like GENI will have to serve an increasing population of users. Serving large number of users would require a large set of available resources. It makes sense to reuse the available infrastructure (hosts, routers, links, etc) as much as possible in order to improve service capacity. However, since an SDN controller is not deployed in a conventional network, testbeds like GENI often have to deploy their own controllers. We have already seen earlier in this chapter that controller is an important component of SDN. We have also seen that deploying multiple controllers is better for the operation of SDN. Now the question is how many controllers do we need and at what locations in a typical SDN to serve a growing population of users. Heller et al. (2012) have used analytical techniques to answer these questions for a 34-node SDN developed by Internet2 in collaboration with several industry partners (Sedmak 2013). It has been shown that the number and location of controllers depend on the:

- Network Topology (which is different for different networks)
- Reaction thresholds (parameters that are specified at the time of design)
- Metric choice (latency, throughput, packet loss, etc).

Since the number and placement of controllers heavily depend on a variety of rapidly changing parameters, a general statement is not sufficient. Using the topology of Internet2 network and communication latency as metrics that are bound by different thresholds, interesting observations have been recorded in (Heller et al. 2012). For example, it has been shown that most medium-sized networks using only one controller can meet the delay thresholds specified by most communication technologies.

3.5 OpenFlow Protocol

After discussing the networking platform in the previous section, we turn our attention to the protocols that enable SDN. A protocol that modifies the table entries of routers based on the instructions from the controller is essential for SDNs. The GENI testbed discussed earlier uses OpenFlow protocol for this purpose, which is discussed in the following.

The OpenFlow protocol enables communication between the controller and programmable routers. The OpenFlow standard is being maintained by the OpenFlow Networking Foundation (ONF), which is a joint initiative of companies like Cisco, Microsoft, Google, etc (Vaughan-Nichols 2011). The first version of OpenFlow, version 1.1, came out in February 2011. OpenFlow 1.1 was then modified in December 2011 as OpenFlow 1.2. ONF has now introduced OpenFlow 1.4. A variety of vendors are dealing with OpenFlow-enabled network elements. To name a few, Cisco, Bell, IBM, Juniper, etc, have shown interest in SDN, and particularly OpenFlow. The architecture of SDN that uses OpenFlow and its operation are discussed in the following.

3.5.1 OpenFlow Network Architecture

As mentioned earlier, OpenFlow protocol runs between the controller and its routers to update their flow tables. The architecture of a network that uses OpenFlow has been shown in Fig. 3.8. In OpenFlow terminology, network routers are referred to as switches. A typical router is responsible for processing the packet headers and forwarding the payload. In SDN, since a router is no longer processing the packet, it is said to be operating as a switch. These switches communicate with the controller on a secure link using the OpenFlow protocol. The secure layer can be provided by protocols such as Secured Socket Layer (SSL) etc. In addition to having a secure channel, these switches have a flow table that is remotely updated by the controller. As can be seen from Fig. 3.8, these switches are connected to the controller on one end, and with the network hosts on the other.

Figure 3.8 shows a software-defined network with a pure OpenFlow switch that has no control on its flow table. Such a switch is known as the dedicated OpenFlow switch. It is also possible to have a switch that has two flow tables: one that caters for production traffic and the other that handles experimental traffic. Only the flow table that handles experimental traffic is updated by the controller. The flow table that forwards normal production traffic will still be updated by the router using conventional protocols. A switch that handles production traffic as well as the experimental traffic

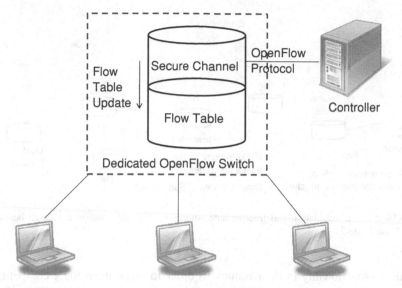

Fig. 3.8 A dedicated OpenFlow switch connected to a controller on one end, and to the network hosts on the other

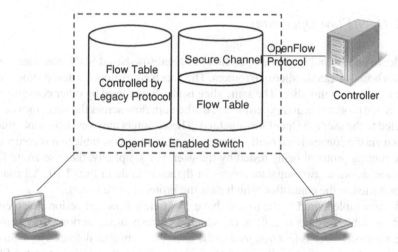

Fig. 3.9 An OpenFlow-enabled switch has two flow tables one of which is programmed by the controller

is known as OpenFlow-enabled switch (McKeown et al. 2008). The architecture of an SDN that uses OpenFlow-enabled switch has been shown in Fig. 3.9. It can be seen from the figure that the switch now has two separate flow tables. The controller is connected to the flow table that handles experimental traffic only. The OpenFlow-enabled switches are preferred over dedicated switches because they can cater for both production and experimental services. Commercial vendors need to add this

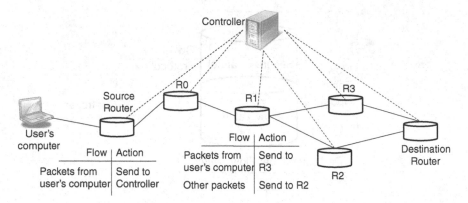

Fig. 3.10 Source, destination and intermediate routers serving information exchange between Node-1 and Node-2

OpenFlow functionality in their routers in order to make them SDN compatible. Various vendors have started modifying their hardware to accommodate OpenFlow.

3.5.2 OpenFlow Operation

In order to understand the basic operation of OpenFlow-based SDN, consider a user who has been allocated a slice of 6 routers. The user has developed a new protocol that has to be tested on this slice. The same slice is also being used by other conventional users. Since production and experimental traffic both flow across the slice, the routers allocated to the user are OpenFlow-enabled. These routers have one flow table that is updated via the conventional routing protocol and another flow table that is controlled by the routing protocol being tested by the user. For simplicity, these separate flow tables are shown as two separate entries in the same table in Fig. 3.10. All routers are connected to the controller, which runs the protocol to be tested.

The flow table stored in the routers have two fields: flow and action. Each entry within the table belongs to a flow, on which a corresponding action is performed. When a network slice is first assigned to the user, the controller defines a *flow*. In our example, a flow constitutes all packets that originate from the user's computer. Flow can also be defined in various other ways as long as it can be distinguished from the rest of the traffic. At the beginning, source router is programmed to send all packets belonging to user's flow to the controller. The other packets are routinely processed by the source router. Upon receiving the first packet, source router forwards it to the controller, which in turn executes the protocol being tested by the user. The controller programs all routers within the slice based on the results computed by executing user's protocol. In other words, the controller sets the flow table entries of all routers within the network after receiving the first packet. This approach is often referred to as the reactive approach. In proactive approach, the controller populates the flow table entries even before the first packet is received (Fernandez 2013).

Once the routers are programmed, controller does not receive the remaining packets of a flow. Note from Fig. 3.10 that the source router can reach destination router using two possible paths: (a) SR-R0-R1-R2-DR and (b) SR-R0-R1-R3-DR. Suppose that the conventional protocol running on this network prefers the first path while the protocol being tested by the user prefers the second. It can be seen from the flow table of R1 (see Fig. 3.10) that this router behaves differently for different packets that move across the network. If R1 intercepts a packet that has been sent by user's computer, it ignores the conventional routing protocol and behaves the way it has been programmed by the controller. In other words, all packets that originate from the user's computer are sent to R3 from R1. The rest are sent to R2. In this way, router R1 serves production traffic as well as the experimental traffic.

3.6 Summary

Software-defined Networking is an innovative trend in communication networks. The main idea behind SDN is to decouple the control and data planes in a router. A router in a typical SDN environment shall only forward data, while all routing decisions shall be made by an external device known as controller. The number of controllers and their placements in the network depends on a variety of factors. However, for medium sized networks, one or two controllers may be sufficient. With rapid popularity of SDN, several networks defined by software have been developed. For example, GENI is an SDN that allows experimentation on live production network. The protocol used in GENI for communication between controller and its routers is known as OpenFlow. OpenFlow is emerging as the most influential protocols of SDN.

This chapter has highlighted the importance of SDN for researchers. It also provides a detailed account of the architecture and operation of networks that use the SDN concept.

References

E. Blanton, S. Chatterjee, S. Gangam, S. Kala, D. Sharma, S. Fahmy, P. Sharma, Design and Implementation of the S3 Monitor Network Measurement Service on GENI, in *4th International Conference on Communication Systems and Networks*, (2012)

Cisco, Software-Defined Networking: Why We Like It and How We Are Building On It. White Paper, (2013)

M. P. Fernandez, Comparing OpenFlow Controller Paradigms Scalability: Reactive and Proactive, in *International Conference on Advanced Information Networking and Applications*, (2013)

FloodLight, Project Floodlight: Open Source Software for building Software-Defined Networks (2014), Available online at http://www.projectfloodlight.org/

GENI, Geni Experiment and Assignment Repository (2013a), Available online at http://groups.geni.net/geni/wiki/GENIExperimenter/ExampleExperiments

GENI, Global Environment for Networking Investigation (2013b), Available online at geni.net.

S. Hares, R. White, Software-defined networks and the interface to the routing system (I2RS). IEEE Internet Comput. **17**(4), 84–88 (2013)

B. Heller, R. Sherwood, N. McKeown, The Controller Placement Problem. ACM Hot SDN (2012)

M. Jarschel, F. Lehrieder, Z. Magyari, R. Pries, A Flexible OpenFlow-Controller Benchmark, in *European Workshop on Software Defined Networking*, (2012)

J.F. Kurose, K.W. Ross, *Computer Networking: A Top-Down Approach* (Addison Wesley, Boston, 2008)

D. Levin, A. Wundsam, B. Heller, N. Handigol, A. Feldmann, Logically Centralized? State Distribution Trade-offs in Software Defined Networks, in *Workshop on Hot Topics in SDN, ACM Sigcomm*, (2012)

N. McKeown, T. Anderson, H. Balakrishnan, G. Parulkar, L. Peterson, J. Rexford, S. Shenker, J. Turner, OpenFlow: enabling innovation in campus network. ACM SigComm. Comput. Commun. Rev. **38**(2), 69–74 (2008)

D. Meyer, The software-defined-networking research group. IEEE Int. Comput. **17**(6), 84–87 (2013)

J.T. Moy, *OSPF: Anatomy of an Internet Routing Protocol* (Addison-Wesley Professional, Reading, 1998)

S. Ortiz, Software- Defined Networking: On the verge of a breakthrough? IEEE Comput. **46**(7), 10–12 (2013)

T. Sedmak, Internet2 and Networking Industry Partners Drive Innovation and Development of SDN and OpenFlow Applications for 100G Network (2013) Available Online at internet2.edu.

S. A. Shah, J. Faiz, M. Farooq, A. Sha?, S. A. Mehdi, An Architectural Evaluation of SDN Controllers, in *IEEE International Communications Conference's Next Generation Networking Symposium*, (2013)

V. Thomas, GENI: Exploring Networks of Future, in *15th GENI Engineering Conference*, (2012)

TraceRoute (2013), Available online at. Network-Tools.com.

S.J. Vaughan-Nichols, OpenFlow: The Next Generation of the Network. IEEE Comput. **44**(8), 13–15 (2011)

X. Yin, S. Huang, S. Wang, D. Wu, Y. Gao, X. Niu, M. Ren, H. Ma, Software defined virtualization platform based on double-FlowVisors in multiple domain networks, in *8th International Conference on Communications and Networking in China*, (2013)

Chapter 4
Opportunistic Networking

4.1 Defining "Opportunity"

All modern communication networks focus on providing high data rates to their end users. Different wireless technologies have a certain value of theoretical maximum data rate that they can offer. Due to a variety of reasons, the actual data rate is always smaller than the theoretical maximum. Let's suppose that a 3 MB file is to be sent over a network that supports 1 MBps. Ideally, this file transfer should take 3 s. Let's further assume, for the reasons explained later in this chapter, that the device sending this file is connected to the network for only 2 s. For a successful file transfer in this case, the device needs to increase its data rate from 1 to 1.5 MBps. Note that this increase in data rate is required not because of large file size. Instead, this increase is required because the connection time is very small. The networking scenarios that typically have small connection times come under the category of opportunistic networking.

Opportunistic communication is the exchange of data whenever an *opportunity* becomes available. Opportunistic networking is a set of scenarios in which overall connection time between the device and the network is the main limiting factor. The amount of information to be sent may be small, however, it needs to be sent quickly because the connection may terminate abruptly. The most fundamental characteristic of opportunistic networking scenarios is that the connections are short-lived and inter-mittent. These short-lived opportunities may arise due to a variety of reasons, some of which have been discussed in this chapter. The reason why these opportunities come and go depends on the networking environment. Some of these environments have been covered in this chapter.

This chapter focuses on a few emerging communication environments that are opportunistic in nature. To be more specific, this chapter explores cognitive radio networks, mobile relay networks and 802.11-based vehicular communication. In addition to providing the basic definitions of these techniques, the main focus is on highlighting their opportunistic nature.

S. F. Hasan, *Emerging Trends in Communication Networks*,
SpringerBriefs in Electrical and Computer Engineering,
DOI: 10.1007/978-3-319-07389-7_4, © The Author(s) 2014

4.2 Opportunistic Channel Access

We have already seen in the previous chapter that network routers can be defined using software. Since routing is a Network layer (layer 3) issue, the previous chapter focused on defining L3 of a network using software. In addition to programming L3, it is also possible to control the physical layer (L1) by means of software. The so-called Software-Defined Radio (SDR) has recently gained popularity, in which a communication device can be programmed to transmit (and receive) on the specified wireless channels (Harris and Lowdermilk 2010). It is well known that wireless channels often face problems like fading, interference, noise, etc. In a typical SDR environment, a device can intelligently detect these problems and instruct its transmitter (and/or receiver) to tune to another wireless channel, if needed. This decision is taken by a code that runs in the background. The technique is referred to as SDR because the radio terminal of a wireless device can be programmed to switch channels intelligently. The following discussion highlights that SDR based techniques are opportunistic in nature.

4.2.1 Cognitive Radio Networks

All communication services are tuned to a certain operating frequency. The transmitter as well as the receiver tune to a specified frequency in order to exchange information. As it stands, the entire usable range of frequencies has been taken up by various technologies (NTIA 2013). In order to accommodate more services and users, we need to explore other parts of the usable spectrum. While 5G cellular technology is examining the use of higher frequency bands to solve this issue, the concept of cognitive radio has also come up as an alternative. Cognitive radio moves a user's transmission session to a frequency that is vacant at that time instant. However, there is no guarantee that the band shall remain vacant for a fixed time interval. This *opportunistic* allocation of frequency applies only to the users that are recognized as secondary users (SU) by the network operators. The primary users (PU) continue to use the spectrum with a higher priority. For example, the TV spectrum bands are allocated to television transmissions. If certain portions of this band are not currently in use by PUs (TV transmitters), they can be used by SUs (communication devices) to exchange information over the network. SUs will have to shift their communication to another frequency as soon as PUs express their intention to use the TV spectrum. Figure 4.1 shows the opportunistic channel access of a SU during the time when that channel is not used by PU. IEEE 802.22 Wireless Regional Area Network (WRAN) is working on standardizing cognitive radio for use in idle TV spectrum bands (Domenico et al. 2012). The idea is to assign different priority levels to SUs and PUs for channel access.

Let's consider Fig. 4.2 to understand the assignment of priority to SUs and PUs in a typical cognitive radio environment. The figure shows two PUs and one SU

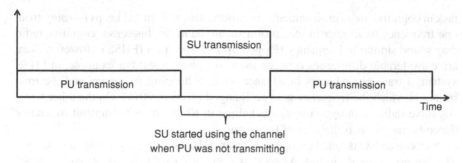

Fig. 4.1 SU accesses the channel when no PU is using it

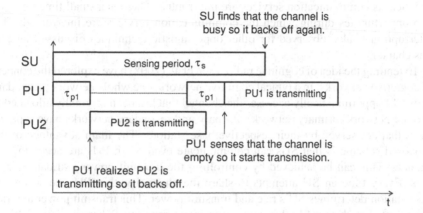

Fig. 4.2 PUs always have a higher priority than SUs

all of which are trying to access the channel together. Channel access privileges among PUs are decided using conventional multiple channel access techniques. For example, 802.11 networks use Carrier Sense Multiple Access/Collision Avoidance (CSMA/CA). In addition to having a fair channel access mechanism, it is possible that PUs maintain priority among themselves (for example using 802.11e). On the other hand, channel access between SUs and PUs is always priority based, such that all PUs have a higher priority. Priority can be set by assigning different channel sensing times. Channel sensing time for SU, τ_s, is set higher than the sensing times of PUs, τ_p. Let's assume $\tau_s > \tau_{p1} > \tau_{p2}$ for the example shown in Fig. 4.2. As can be seen from the figure, PU$_1$ and SU both contend for the channel at the beginning. Since $\tau_s > \tau_{p1}$, PU$_1$'s timer expires first hence it gets access to the channel before SU. Before PU$_1$'s timer expires however, PU$_2$ expresses its intent to use the channel. Since $\tau_{p1} > \tau_{p2}$, PU$_2$ gains access to the channel which forces PU$_1$ to back off. PU$_1$ and SU, while backed off, continue sensing the channel as PU$_2$ uses it for transmission or reception. After PU$_2$, PU$_1$ acquires the channel and SU has to back off once more. It is obvious from this discussion that SU can access the channel only when there is no primary user. Therefore, meeting the QoS requirements of SUs is a challenging task in cognitive radio environments. Secondly, note that an SU keeps *hopping* from one frequency to another in search of a vacant band. In this sense, cognitive radio may sound similar to Frequency Hopped Spread Spectrum (FHSS). However, there are considerable differences between the two technologies. For example, in FHSS system, a transmitter knows in advance its next hopping frequency and the time instant at which the frequency is to be changed (Torrieri 2011). On the other hand, cognitive radio is an opportunistic technique in which the next transmit frequency depends chiefly on availability of the resource.

Suppose an SU is tuned to a certain channel that is originally assigned to several PUs but not currently in use. As soon as a PU arrives, SU backs off and searches for a new vacant channel. During the time that a cognitive node searches for a new channel, its communication services are interrupted. There is a small time between two opportunities during which the communication services are interrupted. This interruption is also observed for other opportunistic techniques discussed later in this chapter.

Extending the idea of cognitive radio, Lien et al. (2008) have explored the concept of cognitive networks. In a typical cognitive network, one whole network (secondary network) opportunistically accesses the channel that has been originally allocated to another network (primary network). Primary and secondary networks have separate users that are served by their respective base stations. The main advantage of the proposed scheme is that SUs can communicate even when PUs are accessing the channel. This can be achieved by controlling the transmit power and data rate of SUs. Every time an SU attempts to share the channel with a PU, the secondary base station determines SU's rate and transmit power. This transmit power and rate are selected such that SU does not cause interference to PU transmission. SU is allowed to access the channel with PU only if such a rate and transmit power can be computed. Otherwise, SU has to back off. Numerical assessment reveals that this method improves the throughput of the secondary users with little impact on the performance of PUs (Lien et al. 2008). Letting SUs communicate simultaneously with PUs also allow some provision of QoS to the secondary users.

4.2.2 Channel Sharing: D2D, M2M and IoT

Opportunistic sharing of a channel with another user, conditioned on a specified rate and transmit power, is also extensively used in Device-to-Device (D2D) communication. The following discussion contains brief comments on D2D setup. A detailed account of D2D communication is given in later in this book.

Networks that allow D2D communication cater for two kinds of users: Cellular User Equipment (CUE) and D2D User Equipment (DUE). CUEs are conventional users that communicate via the base station, while DUEs are cellular nodes that exchange data directly. In order to reuse the frequency band more efficiently, CUE and DUE share the same channel. This is known as the underlay mode of D2D communication. In order that these devices do not interfere with each other, the base station computes the transmit power and permissible data rate of DUEs. Just like cognitive radio networks, DUEs are allowed to communicate only if their transmit power and data rate do not disrupt the communication of CUEs. However, unlike cognitive radio networks, there is only one base station that manages DUEs and CUEs both.

An idea similar to D2D communication is Machine-to-Machine (M2M) communication. Like D2D, M2M also forms an ad hoc network of "machines", which communicate with each other with no (or very little) human intervention (Zheng et al. 2012). M2M communication is gaining rapid popularity with millions of devices expected to participate in the communication process. The number of devices connected to the network may reach 50 billion by 2020 (Ericsson 2011). Accommodating this huge number of devices on an already occupied spectrum is impossible without sharing the spectrum with other services. Note that the information sent in M2M communication is comparatively smaller. This is because machines only have to send signals to notify about the concerned events. Typically, a machine would sense a stimulus and send a message signal to another machine. This message signal of course does not require large bandwidth. The amount of data communicated in M2M is even lower than that required in small-sized file transfer. The use of cognitive radio techniques that allow opportunistic access to the spectrum for machines (as SUs) may be helpful in realizing M2M communication. The M2M devices can act as SUs and share the channel with another network to send data opportunistically.

The concept of M2M communication is one of the main driving forces for the Internet of Things (IoT). While the traditional Internet caters for communication services between humans, IoT is envisaged as a network that allows smart devices to share data autonomously (Palattella et al. 2013). Several applications of IoT are being explored and their feasibility is been examined. It is obvious that the channel sharing shall become more challenging as more and more devices join the IoT. In addition to channel sharing, there are other issues that must be resolved to make IoT a reality. For example, introducing a huge number of sensors and communicating devices would result in considerable electric power consumption. In a world where electricity is already scarcely fulfilling the demand, these sensors will be an additional burden. Therefore, energy efficiency of these devices must be addressed carefully. We have seen earlier in this book that all communicating devices need an address. A million more devices on the network surely requires a million more addresses. Having a well defined address space for devices participating in IoT is also a challenging issue. IPv6 networks with a increased address space can come in handy in this situation. Hurlburt et al. (2012) have analyzed the concept of IoT by critically examining its practical value. It has been concluded that IoT still requires considerable research and development work before it can be introduced.

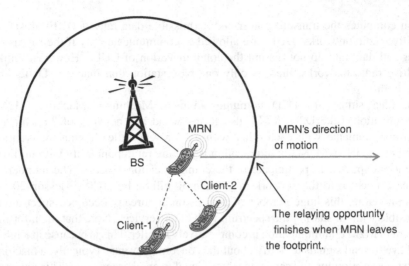

Fig. 4.3 A mobile relay network is opportunistic because relaying service finishes as soon as MRN leaves the BS footprint

4.3 Mobile Relay Networks

The legacy cellular networks have always used a centralized architecture, in which all nodes communicate with each other via the base station. The coverage area of a base station is called its footprint. The geographical area served by a base station is also called a cell. The nodes located at the edge of the cell receive a smaller signal strength from the base station. This leads to a smaller Signal to Noise ratio (SNR) and hence reduced performance at the cell edges. In order to increase the signal strength in these conditions, a relay node is often employed. The use of a relay node is not limited to the cell edges only. In densely populated environments, it is possible to have so-called shadowed region, in which urban obstacles prevent signals from reaching a device. In such scenarios, a relay node is often used to provide a better SNR to the users. A typical relay node acts as a repeater between the nodes and the base station. A relay node, such as one used in LTE networks, is generally static and requires dedicated deployment where ever required (Cox 2012). In order to avoid extra costs of deploying a relay node, Mobile Relay Networks select an ordinary user as a relay node to serve other users.

The mobile relay network shown in Fig. 4.3 is a good example of an opportunistic networking scenario. As the name indicates, the Mobile Relay Node (MRN) is constantly in motion (Stavroulaki et al. 2011). It can provide service to its clients as long as it remains within the footprint of the base station. As it moves out of the coverage area, its services to the clients are suspended. Hence, the communication opportunity in this case is defined as the time during which the mobile relay node remains within the footprint of a base station. As soon as the MRN leaves the footprint, the opportunity for the nodes to communicate is lost. The next opportunity for the edge users arises if another mobile user is selected as new MRN by the base station. Selecting a new node as a relay of course takes time. The communication services for the edge users during this time shall remain suspended. As we noted earlier in this chapter, these interruptions are very common in opportunistic networking scenarios.

4.3.1 Vehicular Relays

The situation becomes more challenging when MRNs are realized for users that are placed inside a vehicle. The nodes that are situated inside the vehicle (i) move at a very fast speed and (ii) have to face the so-called vehicular penetration loss (VPL). The signals from the base station penetrate through the vehicle and lose some of their power. This loss of signal is more significant at the cell edges. Sui et al. (2013) have proposed to place an MRN on the roof of a public vehicle. This MRN provides services to the users traveling on that public vehicle, for example a bus. The use of MRN has been shown to reduce VPL and improve the overall throughput of the users. It is interesting to note that a roof-top MRN serving a set of cellular users inside a vehicle is *not* an opportunistic scenario. This is because MRN shall always stay within the coverage of one base station or move to the realm of another. Even if a user chooses to leave the bus, the cellular services shall still be available from the nearest base station. Therefore, it is reasonable to say that due to the widespread coverage of cellular base stations, cellular networks seldom exhibit opportunistic behavior. However, when vehicles access short-range communication services, they often give rise to opportunistic phenomenon. The next section explores this possibility is more detail.

4.4 Opportunistic Vehicular Communication

With the recent advances in telecommunication electronics, it is now possible to install radio units inside vehicles. These On-board Units (OBU) allow vehicles to exchange information with other vehicles and with the roadside units (RSU). This concept of vehicular communication is getting immense popularity because of its potential in preventing (or at least informing about) possible traffic casualties. The Intelligent Transportation System (ITS) project is underway that explores the use of vehicular communication in a variety of safety and entertainment services (Hasan et al. 2013a). Research and development interest in vehicular communication is driven by the fact that thousands of road casualties are reported every year involving pedestrians and vehicles alike (David and Flach 2010). Communication between vehicles obviously requires an enabling wireless technology. A variety of wireless technologies have been studied and examined for use in vehicular communication applications (Hasan et al. 2013a). Among the popular ones are cellular technology, WiMAX and 802.11 Wi-Fi. IEEE has introduced 802.11p, also known as Wireless Access in Vehicular Environments, (WAVE) for use in vehicular scenarios. 802.11p standard is actually the modified form of the legacy 802.11a standard (Stencil et al. 2007). Enabling vehicular communication over a large area requires deployment of numerous roadside 802.11 Access Points (APs). These APs shall send and receive different safety related messages from the fast moving vehicles. A more economical and hence more preferred approach toward enabling 802.11-based vehicular communication is to use the already available 802.11 APs.

4.4.1 802.11-Based Vehicular Networks

With the massive popularity of 802.11-based Wireless Local Area Networks (WLANs), we have a large population of APs installed inside buildings. The main purpose of 802.11 networks is to provide communication services in indoor low mobility applications. However, their use from outdoor environments in high mobility setups has recently been investigated (Hasan et al. 2010b). In a typical vehicular setup, the roadside APs send and receive information to the vehicles. This kind of vehicular communication is referred to as Vehicle-to-Infrastructure (V2I) communication. V2I communication is useful when a vehicle intends to convey information to the outside networks. For instance, a public bus stuck on a congested road may issue warnings to other vehicles that are not on the same road segment. The bus will have to send a message to the nearest roadside AP, which in turns disseminates the information about traffic congestion everywhere (Hasan et al. 2012a). It is also possible for the vehicles to share information directly with each other like ad hoc networks. This category of vehicular communication is often termed as Vehicle-to-vehicle (V2V) communication (Hasan 2013). V2V communication is best for local sharing of information, such as exchanging warnings related to lane changing, emergency brakes, maintaining inter-vehicular distance, etc. Figure 4.4 shows V2V and V2I communication scenarios. The figure also highlights V2V2I scenario that combines V2V and V2I modes. We argue in the following that both V2V and V2I modes are opportunistic in nature.

The opportunity of communication in V2V setup lasts as long as the vehicles are traveling together in a group. As soon as a vehicle leaves the group, it can no longer exchange information with other vehicular nodes. It is possible for a vehicle to leave one group and join another almost immediately. In this case, there is ideally no interruption in the communication services. Such communication scenarios are known as disruption-free setups, in which the transition from one ad hoc network to another is seamless. However, in most cases, the communication services are disrupted as a vehicle loses one opportunity and gets another. Similarly, communication opportunity in V2I setups exists as long as a vehicle is within the footprint of a roadside AP. After leaving an AP's footprint, vehicle cannot communicate with the network unless it associates with another roadside AP. In typical 802.11-based V2I communication, a disruption period often exists between two opportunities. This is because handing over from one AP to another takes considerable amount of time (Hasan et al.

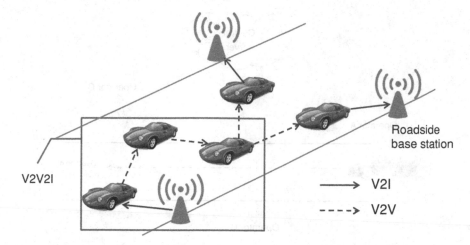

Fig. 4.4 Vehicular communication has two modes: vehicle-to-vehicle and vehicle-to-infrastructure

2013b). During the handover period, all communication services are suspended and the vehicle faces disruption.

4.4.2 Challenges in Opportunistic Vehicular Networks

It has been mentioned in the previous section that opportunistic vehicular communication faces disruption in both V2I and V2V setups. This disruption results in an interrupted communication service for the vehicles. A detailed account of disruption in vehicular communication and its mathematical assessment can be seen in (Hasan et al. 2012b). In V2I setups, this disruption stems from the fact that the deployment of roadside APs is not planned. The deployment of cellular base stations, on the other hand, is planned because they are meant to cover all geographical locations within a particular area. The 802.11 APs, however, are randomly deployed by end users and hence inherently provide disrupted services for highly mobile outdoor users. Therefore, various previous works in this area have focused on *tolerating* disruption rather than removing it altogether (Karkkainen et al. 2013; Eriksson et al. 2008; Hasan et al. 2011).

Vehicles suffer from large handover delays as they make transitions from one AP to another. Reducing this handover time to a seamlessly low value is another important challenge in opportunistic vehicular communication. IEEE has standardized 802.11r (Fast Transition) for reducing the handover delay. However, it is not well suited for fast moving vehicles. Handovers are more difficult to deal with when the previous and new APs belong to different network operators. If a binding between operators, such as one proposed in Wireless Internet Service Provider roaming (WISPr), does not exist, a vehicle may not be entitled to access the next AP despite being within its

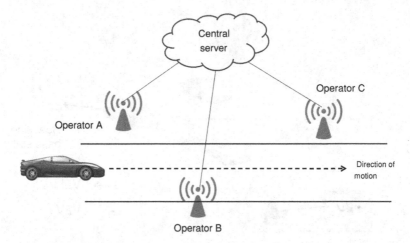

Fig. 4.5 A central server manages the connectivity of a vehicle as it encounters APs operated by different service providers

range (Hasan et al. 2010a). The already sparse deployment of 802.11 APs becomes even sparser as some APs remain unusable for the vehicles because of authorization issues. Therefore, developing a centralized architecture that allows all vehicular users to access all APs belonging to different operators is pivotal for the success of 802.11-based vehicular communication. Figure 4.5 shows such an architecture in which a vehicle can use the APs belonging to three different operators if it has subscribed to only one of them.

4.5 Summary

This chapter covers an interesting attribute of certain emerging networking scenarios. It has been shown in this chapter that these certain networking environments provide short-lived *opportunities* of communication to the end users. These opportunities frequently come and go, making it difficult to provide a desired level of Quality of Service to the users. As examples of opportunistic networking, cognitive radioCognitive radio networks, mobile relay networks and 802.11-based vehicular networks have been discussed. It has been noted that in all these techniques, two consecutive opportunities are generally separated by a disruption period. Communication services during the disruption period remain suspended until a new opportunity arrives.

References

C. Cox, *An Introduction to LTE* (Wiley, Chichester, 2012)

K. David, A. Flach, Car-2-x and pedestrian safety. IEEE Veh. Technol. Mag. **5**(1), 70–76 (2010)

A.D. Domenico, E.C. Strinati, M.-G.D. Benedetto, A survey on MAC strategies for cognitive radio networks. IEEE Commun. Surv. Tutorials **14**(1), 21–44 (2012)

Ericsson, More than 50 Billion Connected Devices. White Paper, (2011)

J. Eriksson, H. Balakrishnan, S. Madden, Cabernet:vehicular content delivery using WiFi, in *ACM MobiCom*, (2008)

F. Harris, W. Lowdermilk, Softwared defined radio: Part 22 in a series of tutorials on instrumentation and measurement. IEEE Instrum. Meas. Mag. **13**(1), 23–32 (2010)

S.F. Hasan, Vehicular communication and sensor networks. IEEE Potentials **32**(4), 30–33 (2013)

S.F. Hasan, N.H. Siddique, S. Chakraborty, On the effectiveness of WISPr in roadside-to-vehicle communications. IEEE Commun. Lett. **14**(9), 818–820 (2010a)

S. F. Hasan, N. H. Siddique, S. Chakraborty, WLAN data rates achievable from roads in low and high mobility environments, in *IEEE International Communications Conference (ICC) Workshop*, (2010b)

S.F. Hasan, N.H. Siddique, S. Chakraborty, Measuring disruption in vehicular communication. IEEE Trans. Veh. Technol. **60**(1), 148–159 (2011)

S.F. Hasan, N.H. Siddique, S. Chakraborty, Extended MULE concept for traffic congestion monitoring. J. Wireless Pers. Commun. **63**(1), 65–82 (2012a)

S.F. Hasan, N.H. Siddique, S. Chakraborty, *Intelligent Transport Systems: 802.11-based Roadside-to-Vehicle Communication* (Springer, New York, 2012b)

S.F. Hasan, N.H. Siddique, S. Chakraborty, Developments and constraints in 802.11 based vehicular communications. J. Wireless Pers. Commun. **69**(4), 1261–1287 (2013a)

S.F. Hasan, N.H. Siddique, S. Chakraborty, Scanning and address allocation delays in vehicular communications. J. Wireless Pers. Commun. **68**(4), 1415–1433 (2013b)

G.F. Hurlburt, J. Voas, K.W. Miller, The internet of things: A reality check. IT Prof. **14**(3), 56–59 (2012)

T. Karkkainen, J. Ott, M. Pitkanen, *Applications in delay-tolerant and opportunistic networks (book chapter)* (The Cutting Edge Directions, Mobile Ad Hoc Networking, 2013)

S.-Y. Lien, C.-C. Tseng, K.-C. Chen, Carrier sensing based multiple access protocols for cognitive radio networks, in *IEEE International Communications Conference (ICC)*, (2008)

NTIA, United States Frequency Allocations. US Department of Commerce, (2013)

M.R. Palattella, N. Accettura, X. Vilajosana, T. Watteyne, L.A. Grieco, G. Boggia, M. Dohler, Standardized protocol stack for the internet of (important) things. IEEE Commun. Surv. Tutorials **15**(3), 1389–1406 (2013)

V. Stavroulaki, K. Tsagkaris, M. Logothetis, A. Georgakopoulos, P. Demestichas, J. Gebert, M. Filo, Opportunistic networks. IEEE Veh. Technol. Mag. **6**(3), 52–59 (2011)

D. Stencil, L. Cheng, B. Henty, F. Bai,Performance of 802.11p waveforms over the vehicle-to-vehicle channel at 5.9ghz. IEEE 802.11 task group p report, (2007)

Y. Sui, J. Vihriala, A. Papadogiannis, M. Sternad, W. Yang, T. Svensson, Moving cells: A promising solution to boost performance for vehicular users. IEEE Commun. Mag. **51**(6), 62–68 (2013)

D. Torrieri, *Principles of Spread Spectrum Communication Systems* (Springer, New York, 2011)

K. Zheng, F. Hu, W. Wang, W. Xiang, M. Dohler, Radio resource allocation in LTE-advanced cellular networks with M2M communication. IEEE Commun. Mag. **50**(7), 184–192 (2012)

Chapter 5
LTE Networks

Wireless communication has brought about many changes in the communication infrastructure. Not only does wireless data exchange simplify the communication process, it also makes room for several innovative applications. Wireless communication became extremely popular on a large scale with the advent of cellular networks in 1980s. Cellular networks primarily cater for voice services (Hasan et al. 2009), however, recent cellular technologies are also increasingly supporting data exchange. The main idea behind cellular networks is that a central node, called the Base Station, serves all wireless nodes within its assigned geographical area. This geographical area is called a cell—hence the name cellular networks. The entire area to be covered, for instance a city, is typically divided into several cells, each of which has its own base station. The coverage area of a typical base station spans about 1 km in radius. The base stations communicate with wireless nodes on one end and with other base stations on the other. Communication between base stations generally takes place over a wired interface. When a node wishes to send information to another node, the information has to be relayed through the base station. If the sending and receiving nodes are not within the same cell, the base station of the sender conveys its data to the base station of the receiver. Thus, the architecture of a cellular network is centralized with base station liaisoning all communication activity. The architecture of a typical cellular network has been shown in Fig. 5.1.

Various cellular technologies have been proposed and used with this architecture over the years. The cellular technology started with the Advanced Mobile Phone System (AMPS) and Nordic Mobile Telephone (NMT) networks that were introduced in 1980s. The freedom from wires and the inherent advantages that come with mobility made these technologies immensely popular. With the passage of time,

S. F. Hasan, *Emerging Trends in Communication Networks*,
SpringerBriefs in Electrical and Computer Engineering,
DOI: 10.1007/978-3-319-07389-7_5, © The Author(s) 2014

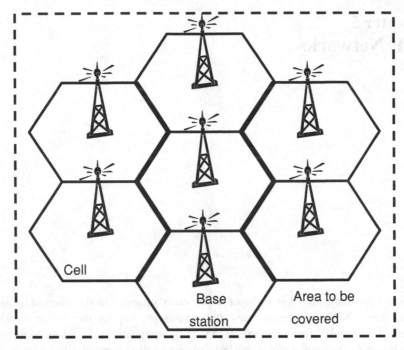

Fig. 5.1 The architecture of a typical cellular network. The area to be covered is divided into *hexagonal cells* each with its own base station

more advanced cellular technologies like Global System for Mobile communication (GSM) and Universal Mobile Telecommunications System (UMTS), etc., were introduced (Huber 2004). Every new cellular technology attempts to improve certain parameters of the previous one. These technologies are classified into several generations of cellular networks. The earliest technologies are referred to as the first generation (1G) cellular networks. AMPS and NMT fall under this category. The technologies introduced later are categorized as 2G, 3G and 4G, such that each generation improves the previous technology in one way or another. Data rate is one of the parameters that are consistently improved by every new technology. By definition, data rate is the amount of information received at the destination node per unit time. It is obvious that supporting high data rates is a must for a cellular technology because higher data rates guarantee quicker file transfers. This chapter discusses the cellular technology that has so far offered the highest data rates to its end users. We explore the Long Term Evolution (LTE) network and the innovation it brings to the cellular infrastructure. The new version of LTE, LTE-Advanced, has also been discussed later in this chapter.

5.1 Background

The LTE network was first introduced by Nippon Telegraph Telephone (NTT) Docomo of Japan. It was later standardized by third Generation Partnership Project (3GPP). 3GPP specifies different standards related to mobile communication services. It has four Technical Specification Groups (TSG) that develop reports and specifications for different components of a communication network (3GPP 2014). The four TSGs active in 3GPP initiatives include Radio Access Networks (RAN), Service and Systems Aspects (SA), Core Network and Terminals (CT), and GSM EDGE Radio Access Networks (GERAN). 3GPP has specified a number of cellular technologies in the past. For example, it specified GSM as a 2G cellular network, High Speed Packet Access (HSPA) and HSPA+ as 3G cellular technologies. The LTE network specified by 3GPP has attracted significant attention in the recent times. This attention has been in terms of research, development and deployment. The main feature of LTE is the provision of data rates that are considerably higher than other cellular technologies. In order to provide high data rates, LTE uses a variety of techniques that are discussed later in this chapter. LTE is an asymmetric network that offers data rates of 300 Mbps in the downlink and 75 Mbps in the uplink with a bandwidth of 25 MHz. It uses Orthogonal Frequency Division Multiple Access (OFDMA) in the downlink and Single-Carrier Frequency Division Multiple Access (SC-FDMA) in the uplink (Pesavento and Mulder 2010). The largest deployment of LTE networks has been witnessed in the United States while South Korea comes in second. LTE is getting popular in other countries too. As of November 2012, 118 commercial LTE networks are already in operation in 54 countries of the world (Report 2012).

It has been mentioned before that the earlier cellular networks primarily cater for voice services with little (or no) support for data. One of the main concerns of LTE networks from the design perspective is that the subscribers are increasingly using data-oriented services. This would mean that LTE has to support data services to succeed as a global cellular technology. Keeping this in view, the architecture of LTE is designed such that it can readily connect to the IP networks that largely support data oriented services. The LTE-IP network convergence is expected to serve all kinds of mobile users. The architecture of LTE network and its interconnection with other communication systems (including the IP network) is discussed in the next section. The second main design consideration is that the mobile traffic has grown, and continues to grow, massively as compared to the previous years. According to Ericsson (2012), mobile traffic will grow 30-fold over the next few years and 100-fold over the next 10 years. The reason for this rise in network traffic is the fact that almost all consumer devices can now connect and communicate over the network. Accommodating this increasing amount of data is one of the main challenges for LTE networks. LTE handles this issue by introducing the concept of D2D communication, which has been discussed in the next chapter.

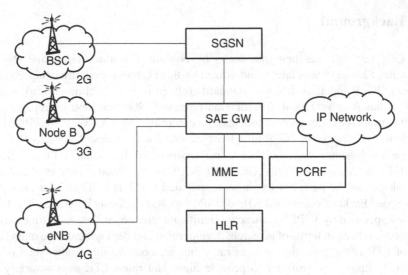

Fig. 5.2 The architecture of an LTE network with connections to other cellular technologies as well as to the IP network

5.2 LTE Network

5.2.1 Network Architecture

The architecture of cellular networks has remained quite the same ever since their inception more than 30 years ago. The LTE network also conforms to the same centralized network architecture. The cell of a typical LTE network has a central base station which serves all clients within the cell. In LTE terminology, the base station is referred to as the evolved Node-B (eNB) while the clients are known as User Equipments (UE). UEs always communicate via their respective eNB while eNBs communicate with each other over wired links. The main components of an LTE network have been shown in Fig. 5.2 (Cisco 2014). The figure shows that LTE networks can be connected with other cellular technologies as well as the IP network.

The System Architecture Evolution Gateway (SAE GW) interconnects different communication technologies with each other. As shown in Fig. 5.2, SAE GW has interfaces to IP network, 2G, 3G and LTE network. The 2G Base Station Controller (BSC) is connected to SAE GW via the Service GPRS Support Node (SGSN). In addition to having interfaces with communication networks, SAE GW also has connections with some of the management entities. For example, it is connected to the Mobility Management Entity (MME) and Policy Charging Rule and Functions (PCRF). As the names indicates, MME is involved in handover processes when UEs change cells. MME is connected to the Home Location Register (HLR) which maintains the current position of a UE. It is pertinent to mention that the position

Fig. 5.3 The lower part of LTE protocol stack showing physical, transport and logical channels

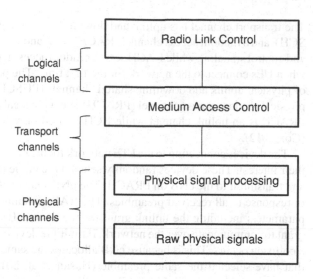

information of a UE is being used in a variety of safety related applications that are not discussed here. HLR also supports MME in managing the handovers. It has been shown in Fig. 5.2 that SGSN and MME are connected via SAE GW. However, a direct connection between the two is also possible. PCRF handles the billing and other account related information (Nuaymi et al. 2012).

5.2.2 Protocol Stack

Like all traditional networks, LTE also has a protocol stack with physical, medium and radio access layers at the bottom and application layers at the top. LTE protocol stack has three main types of channels that interconnect the lower layers. These channels are physical channels, transport channels and logical channels. The physical channels receive raw physical signals from the antenna elements and convey them to the physical processing units. The transport channels provide an interconnection between the physical layer and the MAC layer in an LTE network. Similarly, the connection between MAC layer and Radio Link Control (RLC) is provided by the logical channels. The channels and the layers that they interconnect are shown in Fig. 5.3. The higher layers (IP and TCP etc.,) are connected with RLC via the Packet Data Convergence Protocol (PDCP) layer that has not been shown in Fig. 5.3 for simplicity. Each of the physical, transport and logical channels has various sub-channels. Some of the important sub-channels are briefly discussed in the following.

The logical channel consists of dedicated traffic channel (DTCH), dedicated control channel (DCCH) and common control channel (CCCH). As their names indicate, DCCH and CCCH carry control information while DTCH contains user plane data. DTCH, DCCH and CCCH can be used for uplink and downlink transmissions both.

The transport channel has uplink and downlink shared channels (UL-SCH and DL-SCH), and a random access channel (RACH). UL- and DL-SCH support data transfer and control signaling while RACH serves random access requests only. RACH is used when UEs connect to the network for the first time. The physical channel comprises of physical uplink and downlink shared channels (PUSCH and PDSCH). It also has physical random access channel (PRACH) and a physical broadcast channel (BCH). PRACH is an uplink channel while BCH is used only for downlink transmissions (Cox 2012).

The devices connecting to the LTE network for the first time use PRACH to express their interest. These devices randomly select a preamble (from a set of available 64) and transmit it to the eNB on PRACH. The eNB sends radio access response (RAR) in response to all received preambles. The RAR contains information about several parameters including the uplink grant for the device. Each device uses this uplink grant to communicate over the network. If multiple devices select the same preamble, a collision occurs. This is because eNB allocates the same uplink grant to all devices that have selected the same preamble (Hasan et al. 2013). 3GPP is investigating several methods to avoid this possible collision between the devices.

Channels in an LTE network can also be classified according to the information they carry. In that respect, channels in an LTE network are classified as data and control channels. The data channels contain the user plane data to and from the UE. The downlink control channels contain information from the eNB that is generally related to transmit power control and scheduling commands. The uplink control signals sent by the UEs contain scheduling requests, channel quality indicators (CQI) and Automatic Repeat reQuests (ARQ) acknowledgments. ARQ is a control mechanism that is used to detect errors in data transmissions. Upon successfully receiving a data frame correctly, the receiver sends an acknowledgment packet to the sender. If the sender does not receive an ACK within a certain time limit, it considers the frame as lost and retransmits the same. LTE uses an updated version of ARQ known as the Hybrid Automatic Repeat reQuest (HARQ). HARQ uses the legacy ARQ technique but additionally employs Forward Error Correction (FEC) techniques (Ngo and Hanzo 2014). A few redundant bits are added to the frame in FEC that can correct certain number of errors in the message. Therefore, HARQ requires the sender to repeat the transmission if FEC cannot correct the corrupted message.

5.3 Resource Grid

A resource in LTE networks refer to (a band of) transmission frequency and the time for which it is available for a device. The information about the available resources is maintained in the resource grid. A resource grid can be defined as a combination of available frequencies and time slots that may be assigned to the devices. These time slots are grouped together in frames and are assigned in different configurations. We discuss the frame structure used in LTE networks and the available configurations in the following.

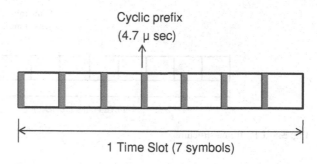

Fig. 5.4 LTE time slot using normal cyclic prefix

Cyclic prefix (4.7 μ sec)

1 Time Slot (7 symbols)

5.3.1 Frame Structure

The smallest unit of transmission time in LTE networks is $T_s = 32.6$ ns. This is the shortest time duration that is of interest to the physical channel processor. A device is generally assigned at least one time slot which is is 15,360 T_s long. This translates to a transmission opportunity of 0.5 ms. Each time slot is itself divided into several symbols. The number of symbols (or resource elements) used to represent a slot depends on the type of cyclic prefix in use. Cyclic prefix helps in reducing the inter-symbol interference. An LTE network can use normal or extended cyclic prefix. A total of 7 symbols represent one slot in case normal cyclic prefix is used. On the other hand, in extended cyclic prefix, one slot is a combination of 6 symbols. The use of normal cyclic prefix is more common in LTE networks mainly because it allows more symbols per slot. The slot structure when normal cyclic prefix is used has been shown in Fig. 5.4. Note that each symbol starts with a prefix whose length is different for normal and extended cycles. The normal cyclic prefix uses a length of 144 T_s which is 4.7 μs long, whereas the extended cyclic prefix uses a length of 512 T_s that is 16.7 μs long. The time duration of each cyclic prefix denote the length of delay spread that can be avoided by the network. For example, the normal cyclic prefix can remove the inter-symbol interference with a delay spread of 4.7 μs. This value is sufficient in most LTE deployments, however, unusually large cells may require a larger prefix. In such cases, extended cyclic prefix is preferred.

Two time slots put together constitutes a sub-frame. Therefore, one sub-frame is 30,720 T_s or 1 ms in duration. One LTE frame is composed of 10 sub-frames. A device that has been assigned one complete frame can use the specified frequency for 10 ms. The frame structure used in LTE is different for different duplexing techniques. Frequency Division Duplexing (FDD) has only one frame structure, as shown in Fig. 5.5. In a typical FDD system, the frequency band allocated to a UE is split into two. One of the resulting sub-bands is used by the UE for uplink transmissions while the other is used by the eNB for downlink transmissions. Since eNB and UE use different frequencies for their respective transmissions, both can transmit simultaneously. Both devices, however, need filters to separate their transmissions from one another. On the other hand, Time Division Duplexed (TDD) systems have 7 different frame structures. These frame structures are referred to as the configurations of the LTE network.

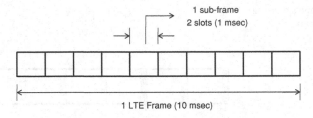

Fig. 5.5 LTE frame structure

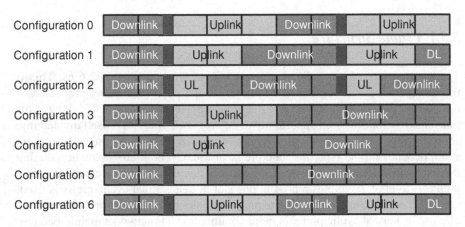

Fig. 5.6 LTE TDD configurations

5.3.2 LTE Configurations

In a typical TDD system, eNB and UE transmit at the same frequency but at different times. One LTE frame is divided into several time slots, some of which are assigned to the eNB while the others are given to the UE. eNB and UE are allowed to transmit in their own time slots only. The number of slots allocated to UE and those allocated to eNB give rise to different configurations. The configurations of an LTE network that uses TDD have been shown in Fig. 5.6 (Cox 2012). The time slots characterized as uplink sub-frame in the figure are the slots assigned to the UE. The slots assigned to eNB are referred to as the downlink sub-frames. It is obvious that different configurations have different time durations for uplink and downlink sub-frames. Note that configuration 0 is symmetric in that both UE and eNB get equal transmission times. In other configurations, for example 4 and 5, eNB gets a larger share of the transmission time. These configurations are particularly useful when eNB has a larger amount of data to send to the UE. This is mostly the case because the traffic sent by the UE is generally smaller in comparison with the traffic sent by eNB. Selecting a suitable configuration for an LTE system is a design issue with no absolute preferences.

Figure 5.6 shows that there is a special sub-frame at the end of UE's (and eNB's) transmission sub-frame. The main task of the special sub-frame is to prevent the transmissions of UE and eNB from interfering with each other. The special sub-frame is similar to the guard band used in Frequency Division Multiple Access (FDMA) systems.

5.4 LTE-Advanced Networks

Different versions of LTE are referred to as its releases. The first version of LTE is characterized as release 8. Various modifications have been proposed in LTE that further enhance its performance. Based on these modifications, various releases of LTE have evolved over the years. The release 10 of LTE, better known as LTE-Advanced (LTE-A), has proposed several interesting modifications in the legacy LTE network. The roll out of LTE-A networks has already started in several countries. The current version of LTE that is being rolled out is release 11. In June 2012, the notion of *Release 12 and beyond* was introduced that focuses on improving energy efficiency and reducing cost (Astely et al. 2013). Among the popular modifications proposed by LTE-A are coordinated multipoint (CoMP) transmission and reception, carrier aggregation and relaying provisions. These techniques improve LTE performance by employing different methods. This section describes the fundamentals of these new ideas one by one. LTE-A also envisages to use Device-to-Device communications that is discussed in the next chapter.

5.4.1 Coordinated Multipoint (CoMP)

Reducing intercell interference is a big issue in almost all cellular networks. UEs that are closer to multiple eNBs have to face significant interference from the neighboring cells. This is particularly true for LTE networks because they use a frequency reuse factor of 1. CoMP has been introduced as a method to reduce intercell interference. The idea was initially introduced in Rel. 11 of LTE, however, more work on CoMP continues in Rel. 12 (Astely et al. 2013).

CoMP allows coordination between several eNBs to serve the UE facing intercell interference. As mentioned earlier, the UEs located at the cell edge generally suffer the most from intercell interference. In CoMP, the serving eNB and interfering eNBs communicate with each other over a high capacity backhaul link to determine parameters like transmission rate, transmit power, scheduling, etc., for the UE (Ghosh et al. 2010). The UE receives data from a set of eNBs such that each eNB conforms to the parameters calculated previously. Since UE receives data from multiple eNBs and not just one, this technique is named coordinated *multipoint*. Figure 5.7 shows that three eNBs use CoMP to serve a UE located at the edge. The serving eNB of the UE is termed as eNB_0. The rest of the eNBs that also contribute to data transfer to

Fig. 5.7 Downlink implementation of CoMP with one
serving and two interfering
eNBs

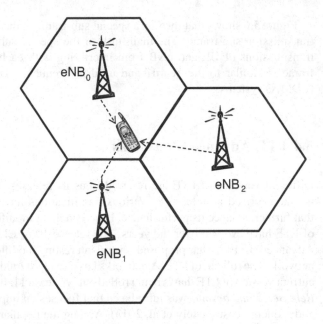

UE are eNB$_1$ and eNB$_2$. Figure 5.7 shows multipoint transmissions in the downlink
direction. Multipoint coordination in the uplink direction is also supported in CoMP
(Ghosh et al. 2010).

Currently, one of the major areas of focus in CoMP is the Channel State Information (CSI) feedback mechanism. CSI is computed using the resources of CoMP
Measurement Set (CMS), which allows the UE to measure rank indicator (RI), pre-
coding matrix indicator (PMI) and channel quality indicator (CQI). In release 10
of LTE, UE computes CSI parameters from one eNB only. However, in the later
releases, UEs have to measure CSI from multiple eNBs. After computing these indicators, UE reports CSI Reference Signals (CSI-RS) to the network (Lee et al. 2012).
Release 11 of LTE is considering the use of CoMP Resource Management (CRM) to
manage CMS of UEs. Using the deployment knowledge of eNBs, the network ini-
tializes CRM. UE measures CRM parameters and reports them back to the network.
Based on this updated knowledge, the network resets the CRM parameters.

5.4.2 Carrier Aggregation

LTE-A network employs the technique called carrier aggregation as a mean to further
increase the data rates. Carrier aggregation relies on the fact that offering a larger
bandwidth to UEs shall always result in higher data rates. Therefore, instead of using
20 MHz bandwidth as used in LTE networks, LTE-A aims to allow the use of up to
100 MHz of bandwidth. The idea of increasing data rates by increasing bandwidth
looks simple but it has one considerable limitation. All network operators do not have

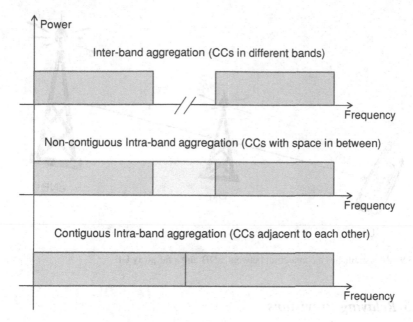

Fig. 5.8 Different types of carrier aggregation methods proposed for LTE-A

a continuous band of 100 MHz to offer to the users. Typically, an operator would have 10 MHz in one band, another 20 MHz in another and so on. Carrier aggregation addresses this problem by allowing a UE to see these scattered bands as one continuous frequency band (Cox 2012). During the Mobile World Congress of 2013, Broadcom demonstrated that two non-contiguous 10 MHz bands can be combined as one to attain 150 Mbps of data rates (Broadcom 2013). Carrier aggregation ideally allows 5 component carriers (CC) of 20 MHz each to be combined together. This results in a 100 MHz band, which would allow downlink data rates of up to 1 Gbps and uplink rates of 500 Mbps. Every UE that wishes to use carrier aggregation has to define its Bandwidth Class, which states the number of aggregated CCs it can handle. It has been pointed out by Ghosh et al. (2010) that peak data rates scale linearly with the number of carriers aggregated together.

There are two ways in which CCs can be combined. Carriers can be aggregated using *inter-band aggregation* and *intra-band aggregation*. In inter-band aggregation, CCs to be combined are located in different frequency bands. This is of course difficult to realize because UEs must be able to support transmission and reception over multiple frequency bands. It also requires well tuned filters. On the other hand, intra-band aggregation combines the frequency bands that lie within the same spectra. The intra-band aggregation can be further classified as non-contiguous and contiguous aggregation. The simplest of these is the contiguous intra-band aggregation in which CCs to be combined are located adjacent to each other. In non-contiguous intra-band aggregation, CCs are in the same frequency band but are separated by a certain spectral space. All three types of carrier aggregation have been shown in Fig. 5.8.

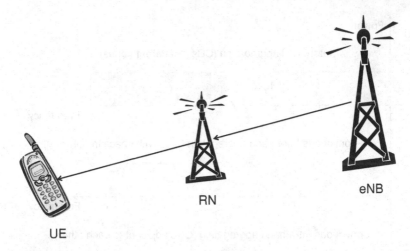

Fig. 5.9 RN establishes a connection between eNB and a far away UE

5.4.3 Relaying Provisions

A relay node (RN) is a device that receives information from one end and transmits the same on the other. It decodes the received signal, re-encodes and re-transmits the same. Relaying achieves higher performance because, unlike amplification, it does not amplify the noise contained in the signal. Relaying is meant to serve several purposes in modern day cellular networks. Firstly, it extends the coverage area of the base stations without requiring additional deployment. Secondly, it provides coverage to the so-called shadowed regions. In dense urban environments there are several areas that do not receive a direct signal from the base station. Deploying a RN closer to such areas can provide coverage which would not be possible otherwise. Finally, RN also helps in increasing the data rates of UEs. It is well known that the Signal to Interference and Noise Ratio (SINR) decreases as the UE moves further away from eNB. According to Shannon's capacity theorem, a decrease in SINR leads to a decrease in data rates. RN located closer to a UE improves its SINR and hence the data rates.

For downlink transmission in LTE, RN takes input from eNB and sends that to a UE as shown in Fig. 5.9. From the view point of an eNB, RNs appear similar to an ordinary UE. RNs' attach procedure to eNBs is also the same as that used by UEs. An eNB that controls one or more RNs is called the Donor eNB (DeNB). There is no difference between eNB and DeNB except the the later controls a relay node. Based on how RNs are used, they can be classified as type 1 or type 2 RNs (Sartori et al. 2011; Yu et al. 2011). Type 1 RN is used to cater for a UE that is located outside the coverage of eNB. This kind of RN has its own physical cell identify (PCI) and is responsible for its own control signaling. Type 2 RNs, on the other hand, are employed within the eNB coverage area to cater for UEs that are located

Fig. 5.10 Block diagram
showing DeNB, eNBs and RN

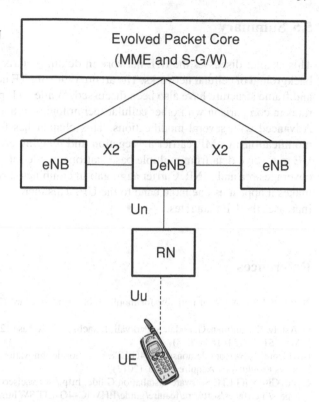

in shadowed areas. Type 2 RNs only perform simple relaying between UE and its concerned eNB. They do not have their own PCIs. The UEs that are served by type 2 RNs are controlled by the eNB itself.

Traditionally, the cellular networks have always maintained a one-hop architecture. In the current LTE network as well, UE and eNB are one-hop away from one another. However, as can be seen from Fig. 5.9, an LTE network that uses RN shall have two-hops (if not more) to and from eNB. The communication interface between DeNB and its RN is termed as the Un interface. There may also be other interfaces between DeNB and its RNs. The interface between RN and its client is referred to as Uu interface. A simplified block diagram of relaying provisions in the LTE network has been shown in Fig. 5.10. The relaying mechanism proposed for LTE networks uses static RNs that are dedicatedly deployed where ever needed. This of course has service limitations in case of any fault that involves the serving RN. An emerging trend is to select an ordinary UE to serve as a RN for the other users. This is a more flexible and low-cost solution because (i) a new RN can be selected as soon as something goes wrong with the existing one and (ii) no dedicated deployment is required. We discuss this idea in more detail in the next chapter.

5.5 Summary

This chapter discusses the LTE network in detail. It starts with a discussion on the background of cellular networks. The architecture of LTE network, its protocol stack and frame structure have also been discussed. While LTE promises the highest data rates in comparison with other cellular technologies, it has been improved as LTE-Advanced with several modifications. This chapter has focused on three of these modifications: CoMP, carrier aggregation and relaying provisions. CoMP allows a UE to receive data from multiple base stations, while relaying improves the eNB's coverage area and SINR. Carrier aggregation combines several frequency bands and makes it appear as one large band to the UE. This increase in bandwidth eventually increases the UE data rates.

References

3GPP, 3GPP – A global initiative for mobile broadband standard (2014), Online at http://www. 3gpp.org/

D. Astely, E. Dahlman, G. Fodor, S. Parkvall, J. Sachs, LTE release 12 and beyond. IEEE Commun. Mag. **51**(7), 154–160 (2013)

Broadcom, Broadcom demonstrates 4G LTE and mobile innovations at Mobile World Congress. News Provided by Acquire Media, (2013)

Cisco, Cisco 4G LTE Software Installation Guide. http://www.cisco.com/c/en/us/td/docs/routers/ access/interfaces/software/feature/guide/EHWIC-4G-LTESW.html Available online, (2014)

C. Cox, *An Introduction to LTE* (Wiley, New York, 2012)

Ericsson, Ericsson Mobility Report. Technical Report, (2012)

A. Ghosh, R. Ratasuk, B. Mondal, N. Mangalvedhe, LTE-advanced: Next-generation wireless broadband technology. IEEE Wirel. Commun. **17**(3), 10–22 (2010)

S. F. Hasan, N. H. Siddique, S. Chakraborty, Femtocell versus WiFi - A Survey and Comparison of Architecture and Performance. Wireless VITAE, (2009)

M. Hasan, E. Hossain, D. Niyato, Random access for machine-to-machine communication in LTE-advanced networks: issues and approaches. IEEE Commun. Mag. **51**(6), 9 (2013)

J.F. Huber, Mobile next-generation networks. IEEE Multimedia **11**(1), 72 (2004)

J. Lee, Y. Kim, H. Lee, B.L. Ng, D. Mazzarese, J. Liu, W. Xiao, Y. Zhou, Coordinated multipoint transmission and reception in LTE-advanced systems. IEEE Commun. Mag. **50**(11), 44–50 (2012)

H.A. Ngo, L. Hanzo, Hybrid automatic-repeat-reQuest systems for cooperative wireless communications. IEEE Commun. Surv. Tutorials **16**(1), 25–45 (2014)

L. Nuaymi, I. Sato, A. Bouabdallah, Improving Radio Resource Usage with Suitable Policy and Charging Control in LTE, in 6th International Conference on *Next Generation Mobile Applications, Services and Technologies*, (2012)

M. Pesavento, W. Mulder, LTE Tutorial (Part-1): LTE Basics. Femto Forum Plenary, (2010)

Report, 4G Mobile Broadband Evolution: Release 10, Release 11 and Beyond. 4G Americas Technical Report, (2012)

P. Sartori, Z. Li, Z. Gong, A. Callard, A. C. K. Soong, LTE relay backhaul design for sparsely-populated environments. in *IEEE GLOBECOM Workshops*, (2011)

Y. Yu, E. Dutkiewicz, X. Huang, M. Mueck, Inter-Cell Interference Coordination for Type I Relay Networks in LTE Systems. in *IEEE Global Telecommunications Conference*, (2011)

Chapter 6
5G Communication Technology

Before LTE technology was introduced, the third generation (3G) of cellular networks were immensely popular . Most countries in the world are still using 3G networks as the complete 4G roll out still requires more time. While LTE network was originally designed as a 4G technology, it could not meet the data rate requirements set aside by the International Telecommunications Union (ITU). The radio communications sector of ITU introduced the International Mobile Telecommunications Advanced (IMT-Advanced) as a set of specifications that defines 4G networks. For high mobility environments, 4G technology must offer the peak data rates are set to 100 Mbps and 1 Gbps in the stationary conditions (Freedman 2009). Since the earliest release of LTE could not meet these requirements, it is not considered as a 4G technology. It is worth noting that LTE is not the only technology that is aiming to become 4G technology. The Mobile WiMAX, first introduced in South Korea in 2006, is also improving its earlier release to enter the 4G domain (Portela and Diaz 2013). LTE technology, much like Mobile WiMAX, still needs considerable modifications (Lahetkangas et al. 2012), some of which have been discussed in the previous chapter under the umbrella of LTE-A.

This chapter explores the Fifth Generation (5G) of communication networks. While considerable work is in progress in beyond-3G technologies, the interest in 5G is beginning to grow gradually. Most of the work is still in the research phrase. All previous versions of the cellular networks (as well as other communication networks) operate in the sub 3 GHz band. Because of the overwhelming use of 700 MHz to 2.6 GHz spectrum, there exists no room to accommodate future wireless technologies. Therefore, 5G technology sets out to use the frequency band that spans from 3 to 300 GHz. This chapter discusses the 5G idea, its associated issues and possible solutions. We show that device-to-device communication shall play a key role in implementing 5G technology.

This chapter starts by outlining the expectations from 5G cellular technology. It also points out the limitations faced by 5G and their possible solutions. New concepts like UE Relaying and Device-to-Device communication are discussed at

S. F. Hasan, *Emerging Trends in Communication Networks*,
SpringerBriefs in Electrical and Computer Engineering,
DOI: 10.1007/978-3-319-07389-7_6, © The Author(s) 2014

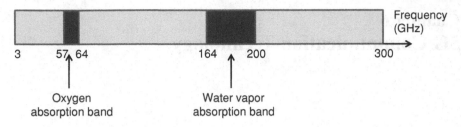

Fig. 6.1 Frequency band above 3 GHz band

length. While discussing D2D communication, various interesting research issues are also highlighted.

6.1 5G Expectations and Limitations

Within the available high frequency band that goes from 3 to 300 GHz, at least 252 GHz is available for use (Pi and Khan 2011). Previously, Local Multipoint Distributed Services (LMDS) were using some of this band. However, their use has now been discontinued. 5G cellular technology is currently being considered for use in 28 and 36 GHz bands. These frequency bands are referred to as the millimeter band (mm-band). Consequently, 5G communication is often termed as millimeter-wave communication. The communication is called mm-wave because the wavelength of the radio waves in this frequency range is in the order of millimeters. Figure 6.1 shows the high frequency band and the available spectrum (Pi and Khan 2011).

In general, a 5G network is at least expected to:

- use larger frequency spectrum at higher frequencies,
- accommodate more users on the network,
- offer much higher data rates, and
- use highly directional antennae.

It is obvious that large frequency band can accommodate more users and can provide higher data rates to each user. The use of directional antennae becomes a necessity because the radio waves in mm-wave band cannot travel a long distance. Directional antennae are meant to help by focusing the signal power in a desired direction.

While the idea of using high frequency band for transmission is reasonable in terms of the available spectrum, it has its own complications and limitations. The biggest limitation of using 28 and 36 GHz bands is that the radio waves cannot travel long distances. This implies that the base stations as well as the mobile nodes shall have small transmission range. It is well known that the penetration loss also increases with the increasing transmission frequency. Therefore, radio waves at 5G frequencies shall have difficulty penetrating through buildings and other urban infrastructure. Moreover, it has been shown in (Rappaport et al. 2013) that seasonal variations also adversely affect the radio waves that are tuned at such high frequencies. Among these issues, the problem of low penetration and smaller coverage area are the most significant ones. The research community is addressing these problems by examining various interesting solutions.

In order to get rid of the problems associated with high penetration loss, the use of femtocells is being considered. Femtocell base stations are deployed inside buildings and hence provide a high received signal strength to the indoor UEs. The mini base stations are connected with the backbone LTE network over a wired interface. When an indoor UE receives a signal from LTE-controlled femtocell base station, the signal gets to penetrate through the building (Xenakis et al. 2014). This is because the femtocell base station is already located inside the building. On the other hand, in order to address the coverage area issue, we have seen in the previous chapter that RNs can increase the coverage region of an eNB. Excessive use of RNs can increase the coverage area but shall incur considerable installation and maintenance cost. Instead of using RNs, another alternate is to deploy dedicated 5G base stations and have them work with the legacy 4G base stations. Based on this idea, Pi and Khan (2011) have proposed the so-called mm-wave Mobile Broadband (MMB) architecture. The proposed network architecture comprises of legacy 4G base stations that are separated by a distance of 1 km, and 5G base stations that are 500 m apart. It has been proposed that 4G base stations take care of all control signaling while 5G base stations handle the data exchange between UEs. This idea also requires considerable dedicated deployment to accommodate 5G services. The following section discusses an innovative method of extending the range of 5G enabled eNBs using mobile relays.

6.2 UE Relaying

The concept of UE-relaying has recently been introduced (3GPP 2013), in which ordinary UEs are selected to serve as RNs for other users. This method is regarded as an efficient and cost effective way for the provision of relaying services because it does not require dedicated deployment. Since ordinary UEs are mobile, it is possible that a UE currently serving as RN moves out of the coverage area and attaches to a new eNB. In this case, a new UE is selected to replace the previous one. There are at least three important design considerations for realizing an effective UE-relaying mechanism. The first is related to the security of the entire network. Since the UE selected to serve as RN, referred to as RUE, shall be forwarding traffic from other UEs, it is important to make sure that it is authorized and secure. The second important consideration is the time delay in selecting a new RUE as the old one moves away from the current eNB. Until a new RUE is selected, all UEs that were receiving service from the previous RUE shall remain stranded. Therefore, the new RUE should be selected with minimum time delay. Finally, the UE selected as RUE shall consume more energy because of relaying data of other UEs. Developing an energy efficient relaying mechanism is pivotal for long term operation of RUEs.

UE-relaying can be classified into two categories: UE-to-Network Relaying and UE-to-UE Relaying (3GPP 2013). Both types of UE-relaying have been shown in

Fig. 6.2 UE-to-network and
UE-to-UE relaying mecha-
nisms

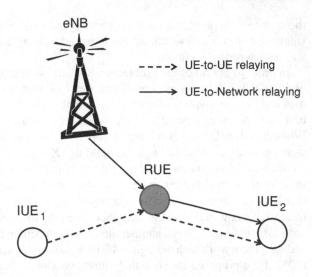

Fig. 6.2. In UE-to-Network relaying, RUE is connected to its serving eNB on one end, and with the served UE on the other. This type is typically employed in cases where certain UEs are outside eNB's coverage area. Such UEs are referred to as the Isolated UEs (IUEs). Thus, the UE-to-Network relaying is mainly used to extend the coverage of eNBs. The UE-to-UE relaying, on the other hand, has been designed to support Proximity Services (ProSe) as envisaged for LTE-A networks. In fact, the National Public Safety Telecommunications Council (NPSTC) has proposed to use LTE network for the provision of ProSe (NPSTC 2013). ProSe is meant to provide quick data exchange between nearby UEs under emergency conditions. In an unprecedented event that affects the usual operations of eNB, ProSe is supposed to provide an eNB-free communication path between UEs. Figure 6.2 shows that IUE_1 can reach IUE_2 through RUE, such that eNB is not involved in the communication process.

Recall from our discussion on LTE in the previous chapter that all communication activity within a cellular network has to pass through eNB. On the other hand, Fig. 6.2 shows that the communication path between RUE and ordinary UEs is a direct cellular link with no involvement of eNB. This type of data exchange between two UEs that is independent of eNB is known as Device-to-Device (D2D) communication. The following section discusses D2D communication in detail.

6.3 Device-to-Device Communication

The modern day communication networks are dealing with an enormous amount of network traffic, which is set to increase in the coming years (Ericsson 2012). Because the eNB relays entire network traffic originating from all UEs within its coverage region, it can face significant congestion and result in a bottleneck. An emerging idea

Fig. 6.3 An LTE cell with a
CUE and a D2D pair

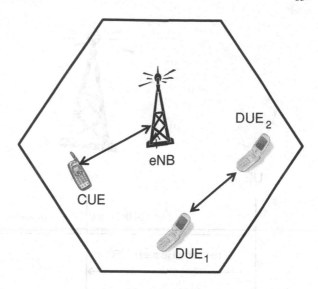

is to off load the relay of traffic through eNB by letting UEs communicate directly. The so-called Device-to-Device communication allows two UEs to exchange information on direct D2D links, bypassing the eNB. All UEs that use D2D communication are referred to as D2D UEs or DUEs. Therefore, an LTE network that supports D2D communication has two kinds of UEs. One is the DUEs, as explained above, and the other is the conventional cellular UEs (CUEs) that communicate through the eNB. Figure 6.3 shows an LTE cell that accommodates a CUE and two DUEs. Two DUEs are collectively referred to as a D2D pair. A D2D pair is composed of a DUE transmitter (DUE_T) and a DUE receiver (DUE_R). It has been mentioned previously that data rates is an important parameter that reflects on the overall performance of the network. In an LTE network that uses CUEs and DUEs both, the collective data rate of CUEs and DUEs is considered as the main performance indicator. The collective data rate of CUEs and DUEs is referred to as the sum-rate.

While eNB does not act as a relay for the participating DUEs, it plays an active role in setting up their connection. After a D2D connection is established eNB is taken out of the picture. A detailed account of cellular D2D connection setup can be found in Lei et al. (2012). A brief summary has been given in the following and its graphical representation has been shown in Fig. 6.4.

Two reasonably close UEs identify each other using Radio Network Temporary Identifier (C-RNTI) and Channel Quality Indicator (CQI) reports. After identifying each other, UE_1 sends a Scheduling Request (SR) to eNB. SR indicates that UE_1 wishes to convey a message to eNB. In response, eNB allocates limited uplink resources to UE_1 to receive its message. Using these resources, UE_1 sends Buffer Status Report (BSR) to eNB as a formal expression of interest to send data to UE_2. Taking into account the CQI reports sent by UE_1 and UE_2, eNB reserves suitable resources for D2D communication. The final decision is sent to UE_1 and UE_2 on

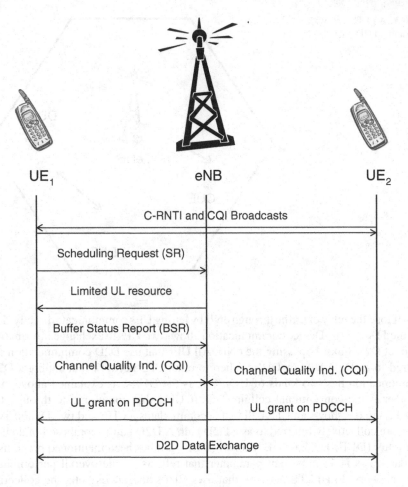

Fig. 6.4 Connection setup between two DUEs in D2D communication

Physical Downlink Control Channel (PDCCH). The two UEs, henceforth DUEs, can now commence direct data exchange without any help from eNB. The eNB reserves the radio resource for D2D communication depending on the mode of operation, as discussed in the following.

6.3.1 Modes of Operation

An LTE network is assigned with a range of frequencies that is used by eNB and its CUEs for data transmission. The manner in which frequency is allocated to DUEs gives rise to two modes of operations of the LTE network. An LTE network that supports D2D communication can operate in either the underlay or overlay modes (Doppler et al. 2009). In the overlay mode, the overall frequency band of an LTE network is divided into two parts and assigned to CUEs and DUEs. Since a portion of the band that was originally assigned to CUEs is given away to DUEs, CUEs have to tolerate a certain decrease in their data rates. On the other hand, in the overlay mode, CUEs and DUEs do not interfere with each other because both of them operate in separate frequency bands. In the underlay mode, both DUEs and CUEs operate on the same frequency band. This mode is better in terms of frequency reuse and spectral efficiency, however, it may lead to considerable interference between CUEs and DUEs that are reusing the same frequency. The network operators generally prefer the underlay mode because reusing the same frequency can accommodate more UEs in a cell. In order that CUEs and DUEs operate on the same frequency band without interfering with each other, their transmit power and resources must be controlled in an intelligent manner. Note that CUEs and DUEs can reuse both uplink and downlink resources. However, it is better that they share the uplink resources only. The downlinks always contain transmissions from eNB that typically have very high power. Controlling interference between eNB and DUEs is considerably difficult on the downlink where eNB transmits at a high power.

6.3.2 Other D2D Technologies: Wi-Fi Direct

Our discussion has so far covered the cellular D2D communication, however, other networking technologies have also evolved that also employ D2D concept. We briefly highlight one of these technologies in the following for completeness. The main focus continues to remain on the cellular D2D communication in the rest of this chapter.

Wi-Fi Direct is a fresh idea that is being examined in the research community. As its name indicates, it is an evolved version of Wi-Fi, or more specifically, IEEE 802.11 Wireless Local Area Networks (WLANs). A conventional Wi-Fi network is meant to provide network connectivity over a smaller geographical expanse typically in the indoor environments (Rackley 2007). Wi-Fi networks have two modes of operation. One is the ad hoc mode in which all 802.11-complaint nodes exchange data directly with each other. There is no base station to mediate the communication activity so this idea sounds similar to D2D communication. However, unlike D2D communication, multi-hop communication is allowed in 802.11 ad hoc networks.

The second mode of operation in Wi-Fi networks is the infrastructure mode. A central entity called the Access Point (AP) handles all communication activity in 802.11-based infrastructure network. This is similar to the conventional cellular networks where a central base station manages all communication activity. 802.11 infrastructure networks are used whenever a Wi-Fi wants to access the Internet. The ad hoc mode only suffices for communication over a smaller range. More recently, Wi-Fi Direct has been introduced as another mode of operation of Wi-Fi Networks. Wi-Fi Direct is different from the other modes because a node can act both as a client or as an AP (Mur et al. 2013). Any node can assume the role of an AP and relay

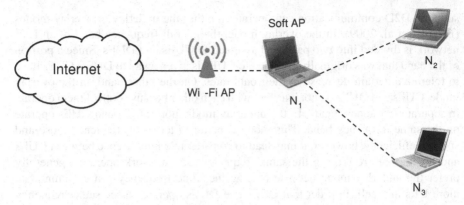

Fig. 6.5 A sample use case scenario of Wi-Fi Direct

information to and from its neighboring nodes. Such an AP is often referred to as the Soft AP. It follows that the nodes in Wi-Fi Direct have to implement both client and AP functionalities. The role assumed by each device is decided dynamically. Figure 6.5 shows a typical use case scenario of Wi-Fi Direct. It has been shown that Wi-Fi AP provides network services to a node N_1. Node N_1 then assumes the role of AP and serves two other nodes N_2 and N_3. Upon careful inspection, it is obvious that Fig. 6.5 has distinct similarities with UE Relaying shown in Fig. 6.2.

6.4 D2D Communication Underlaying LTE-A

In order to underlay D2D communication in LTE-A networks, the transmit power and resource allocation for CUEs and DUEs must be carefully handled. Several methods for transmit power control and resource allocation have been examined in literature. Some fundamental ideas are discussed in the following with suitable analogies to the real world.

6.4.1 Transmit Power Control

Imagine that two persons who wish to speak to their respective colleagues in the same room at the same time. Both of them must speak at a reasonable volume level so that both can convey their message to their colleague without interfering with the speech of the second person. The level of volume in human speech is analogous to the transmit power used by the devices in wireless networks. All wireless devices transmit at a certain power level, just like humans speak at a certain level of volume. Like other wireless networks, LTE-A puts a limit on the transmit power used by its UEs. The transmit power of a CUE is given in Eq. 6.1 (3GPP 2009).

$$p_c = min\left\{p_{max}, \quad p_0 + 10 \cdot \log_{10} M + k \cdot PL_c\right\}, \quad\quad\quad (6.1)$$

where p_{max} is the maximum allowed transmit power of the CUE, p_0 is the UE specific base power, M is the number of resource blocks, k is the path loss compensation factor determined by eNB and PL_c is the high-layer filtered path-loss between the eNB and the CUE.

On one hand, a higher transmit power increases reachability, but on the other hand, it interferes with the transmission of the neighboring UEs that are reusing the same frequency. In a network that uses CUEs and DUEs both, effective transmit power control mechanisms should be in place so that both can communicate with their intended receivers without problem.

The transmit power of DUEs has not been standardized yet. There have been several research works that focus on selecting a suitable power level for the DUEs. Janis et al. (2009b) propose that eNB should select the transmit power of DUEs based on the power levels currently used by the CUEs. This method has been shown to improve system performance particularly when a CUE and a D2D pair reuse the uplink resources. However, this scheme has been tested for one cell with one eNB, one CUE and one D2D pair. It is also possible to assign different priority levels to CUEs and DUEs while controlling their transmit power. Yu et al. (2009) first discuss a general scheme that does not use priority, and then compare the resulting sum-rates with an optimized scheme that prioritizes CUEs over DUEs. In order to ensure that priority is given to CUEs, their SINR level is monitored continuously. If the SINR of CUE drops below a specified threshold, the transmit power of DUE reusing the same resource has to be reduced. As far as a user is concerned, it is not important whether its data is relayed through eNB or not. All that matters is the quality of service (QoS). If both CUEs and DUEs are regular subscribers of the network, prioritizing one over another cannot deliver the same QoS to both. Therefore, exploring the benefits of a fair transmit power control mechanism may be an interesting future direction.

6.4.2 Resource Allocation

Imagine that two professors need to deliver lectures to their respective students at the same time. If both are allowed to speak in the same lecture hall, none of them shall be able to deliver the lecture contents effectively. Both professors are assigned a different lecture hall. In wireless networks, two nodes that wish to send data at the same time are assigned different frequencies. It has been pointed out before that assigning different frequencies to CUEs and DUEs (overlay mode) wastes bandwidth. Therefore, in underlaying D2D communication, a CUE and a D2D pair reuse the frequency that causes minimum interference.

Interference Tracing (IT) and Tolerable Interference Broadcasting (TIB) schemes for resource allocation have been proposed by Peng et al. (2009). In IT scheme, the main focus is on determining the CUEs that would cause minimum interference to DUEs when uplink resources are shared. The eNB broadcasts the Radio Resource

Management (RRM) messages that contain the information on uplink resource allocation. Using this information, DUEs determine the CUEs which are expected to cause interference that is lower than a specified threshold. On the other hand, TIB scheme aims at preventing DUEs from interfering with the eNB. The eNB estimates and broadcasts the tolerable interference for each radio resource. DUEs select only those resources that cause smaller interference to eNB than the advertised value.

Instead of dealing with transmit power control and resource allocation as two separate issues, several works have combined the two mechanisms together. For instance, Janis et al. (2009a) have proposed a combined transmit power and resource allocation mechanism. Interestingly, it allows DUEs to transmit data on the downlink resources as well. However, only those downlink resources are allowed that contain eNB's transmission destined to a CUE which is located far away from the DUE. In addition to employing this resource allocation mechanism, a transmit power control has also been specified (Janis et al. 2009b). The transmit power of CUEs and DUEs is selected such that the minimum SINR of the CUEs stays higher than a specified threshold. Gu et al. (2013), on the other hand, propose a combined scheme that focuses on increasing the overall sum-rate of the network. The eNB first assigns the resources to CUEs. The resource allocated to a CUE is then assigned to a D2D pair that maximizes the overall sum-rate. Depending on the thresholds set aside for CUEs and DUEs, the transmit power of DUEs is controlled to ensure high sum-rate.

So far in our discussion, D2D users reuse the same cellular frequencies intelligently. However, it is also possible to offload the D2D users to some other spectrum. For example, Pyattaevy et al. (2013) propose that D2D pair should be offloaded to Wi-Fi frequencies. While this idea may reduce interference between CUEs and DUEs, it may end up reducing the QoS of DUEs. This is because the Wi-Fi band is already in use by several other services.

6.5 Summary

While 4G networks are being rolled out in many countries across the world, the research work in 5G technology is steadily gaining momentum. The currently used frequency spectrum is now completely in use, leaving no spectral space for future wireless technologies and services. 5G technology addresses this issue by examining the possibility of using higher frequencies of data transmission. While the high frequency band promises high data rates and network capacity, the range and penetration of radio waves are quite limited. UE relaying has been recently introduced as a method to increase the coverage of eNB if it uses the 5G technology. In order to successfully implement UE relaying, the UEs in an LTE network must be able to communicate directly with each other. To cater for this issue, the so-called Device-to-Device communication has been introduced and discussed in detail.

References

3GPP, Study on architecture enhancements to support proximity services (prose). 3GPP Specification (June), (2013)

3GPP, Evolved Universal Terrestrial Radio Access (E-UTRA): Physical Layer Procedures. Technical Specification, (2009)

NPSTC, Public Safety Broadband: Push-to-Talk over Long Term Evolution Requirements. NPSTC Public Safety Communications Report, (2013)

D.C. Mur, A.G. Saavedra, P. SERRANO, Device-to-device communications with WiFi direct: overview and experimentation. IEEE Wirel. Commun. **20**(3), 96–104 (2013)

K. Doppler, M. Rinne, C. Wijting, C.B. Ribeiro, K. Hugl, Device-to-device communication as an underlay to LTE-advanced networks. IEEE Commun. Mag. **47**(12), 42–49 (2009)

Ericsson, Ericsson Mobility Report. Technical Report, (2012)

A. Freedman, International Mobile Telecommunications-Advanced, in *IEEE International Conference on Microwaves, Communications, Antennas and Electronics Systems*, (2009)

J. Gu, S.J. Bae, S.F. Hasan, M.Y. Chung, A combined power control and resource allocation scheme for D2D communication underlaying an LTE-advanced system. IEICE Trans. Commun. **96-B**(10), 2683–2692 (2013)

P. Janis, V. Koivunen, C. Ribeiro, J. Korhonen, K. Doppler, K. Hugl, Interference-aware resource allocation for device-to-device radio underlaying cellular networks, in *IEEE Vehicular Technology Conference*, (2009a)

P. Janis, C.-H. Yu, K. Doppler, C.B. Ribeiro, C. Witjing, K. Hugl, O. Trikkonen, V. Koivumen, Device-to-device communication underlaying cellular communication systems. Int. J. Commun. Netw. Syst. Sci. **2**(3), 169–178 (2009b)

E. Lahetkangas, K. Pajukoski, E. Tiirola, J. Hamalainen, Z. Zheng, On the performance of LTE-Advanced MIMO: How to set and reach beyond 4G targets, in *18th European Wireless Conference*, (2012)

L. Lei, Z. Zhong, C. Lin, X. Shen, Operator controlled device-to-device communications in lte-advanced networks. IEEE Wirel. Commun. **19**(3), 96–104 (2012)

T. Peng, Q. Lu, H. Wang, S. Xu, W. Wang, Interference avoidance mechanisms in the hybrid cellular and device-to-device systems, in *IEEE International Symposium on Personal, Indoor and Mobile Radio Communications (PIMRC)*, (2009)

Z. Pi, F. Khan, An introduction to millimeter-wave mobile broadband systems. IEEE Commun. Mag. **49**(6), 101–107 (2011)

N.A.P. Portela, B.R. Diaz, Performance comparison between the air interfaces of LTE and mobile WiMAX. IEEE Latin America Trans. **11**(4), 1001–1006 (2013)

A. Pyattaevy, K. Johnsson, S. Andreevy, Y. Koucheryavyy, 3GPP LTE Traffic Offloading onto WiFi Direct Traffic Modeling, Subscriber Perception Analysis and Traffic-aware Network, in *IEEE WCNC Workshop on Mobile Internet*, (2013)

S. Rackley, *Wireless Networking Technology: From Principles to Successful Implementation* (Elsevier, Newnes, 2007)

T.S. Rappaport, S. Sun, R. Mayzus, H. Zhao, Y. Azar, K. Wang, G.N. Wong, J.K. Schulz, M. Samimi, F. Gutierrez, Millimeter wave mobile communications for 5G cellular: it will work. IEEE Access. **1**, 335–349 (2013)

D. Xenakis, N. Passas, L. Merakos, C. Verikoukis, Mobility management for femtocells in LTE-advanced: key aspects and survey of handover decision algorithms. IEEE Commun. Surv. Tutorials. **16**(1), 64–91 (2014)

C.-H. Yu, O. Tirkkonen, K. Doppler, C. B. Ribeiro, Power optimization of device-to-device communication underlaying cellular communication, in *IEEE Conference on Communications (ICC)*, (2009)

Chapter 7
Executive Summary

This book has covered some of the most interesting and recent trends pertinent to communication networks. The topics covered in this book may be classified into two broad categories. The first is a set of topics that is related to IP net- works. The second set of topics covered in this book is related to the cellular technology. Brief highlights of the topics discussed in both sections are given in the following.

7.1 IP Networks

IP networks use packet switching and provide network access to millions of users worldwide. The backbone of the present day packet switched networks is being modified by introducing IP version 6. IPv6 networks improve the routing speed by simplifying packet headers. More importantly, IPv6 networks increase the address space thus allowing more users to access the network services. The IPv4 networks, which are still widely used, cannot support an increasing number of users and hence need to be replaced. While IPv6 improve routing speed and address space, the notion of software-defined networking has been introduced to make networks more flexi- ble. The networks defined by software are programmable, in that their response to incoming packets may be changed dynamically. A popular SDN called GENI and its protocol OpenFlow have been discussed in detail. The importance of SDN in improving research methods has also been highlighted in Chap. 3. This book also covers the networking environments that are referred to as opportunistic. In this con- text, issues like cognitive radio, mobile relay networks and vehicular communication have been discussed. In addition to covering the basics of these emerging ideas, their opportunistic nature has been carefully examined.

S. F. Hasan, *Emerging Trends in Communication Networks*,   71
SpringerBriefs in Electrical and Computer Engineering,
DOI: 10.1007/978-3-319-07389-7_7, © The Author(s) 2014

7.2 Cellular Networks

In addition to IP networks, this book also discusses the recent developments in cellular technology. It presents a detailed account of LTE networks, which are becoming increasingly popular in terms of research, development and deployment. The discussion on LTE starts with a quick look at fundamental design considerations, followed by an examination of its protocol stack, architecture and resource grid. While LTE networks are currently in the deployment stage, research is being carried out on LTE-Advanced (LTE-A) networks. This book has covered some of the modifications proposed in LTE-A, including coordinated multipoint, carrier aggregation and relaying provisions. LTE networks, like most wireless networks, use the sub-3 GHz band for transmissions. This band is now over populated with wireless services and cannot support any new deployments. Motivated mainly by this fact, the idea of 5G networks is getting mature which aims at using the higher frequency band for transmissions. These networks are being envisaged for 28 and 36 GHz bands, which are expected to provide very high data rates. However, the wireless propagation characteristics at high frequencies are not suitable for long range transmissions. The concept of UE Relaying has been discussed in this book that targets range extension using mobile relay nodes. In order to realize UE Relaying, the notion of Device-to-Device communication has also been covered in considerable depth.

Appendix A
Exercise Questions

You shall now be able to answer the following questions:

- Explain how packet-switched networks are suitable for supporting bursty traffic?
- What are the two main modifications in IPv6 in comparison with IPv4?
- Write the subnet prefix of the following
  $2022 : A152 : 5EF5 : 79FD : 24C8 : 085A : 8C6E : 6254/64$.
- What is the interface ID of a device with the following MAC address
  $E8 - 11 - 32 - 34 - 51 - B5$?
- Explain how the following processes proceed in IPv6 network: (i) stateless auto-configuration and (ii) DNS configuration
- What is an anycast address and in what situations it cannot be used?
- Explain the five messages supported by Neighbor Discovery Protocol (NDP).
- In Domain Name System (DNS), what is the difference between RDNSS and DNSSL?
- Explain the flow label field in IPv6 packet header.
- How extension headers reduce the computational burden on intermediate routers?
- What do you understand by separating the control and data planes?
- What do you understand by a logically centralized architecture?
- Name at least three architectural designs that are being considered for software-defined networking.
- Explain how software-defined networking can be useful for researchers?
- What changes does a controller make while programming a router?
- What is the use of OpenFlow Protocol in a software-defined network?
- Differentiate between OpenFlow-enabled and dedicated OpenFlow switches.
- Why is a router referred to as a switch in OpenFlow architecture?
- What is a floodlight controller? How many other controllers can you name?
- How is opportunistic networking different than the conventional networking paradigms?
- Why the provision of quality of service a big challenge in opportunistic networks?

S. F. Hasan, *Emerging Trends in Communication Networks*,
SpringerBriefs in Electrical and Computer Engineering,
DOI: 10.1007/978-3-319-07389-7, © The Author(s) 2014

- With the help of a diagram, show the channel access mechanism in a cognitive radio network with nodes PU_1, PU_2, SU_1 and SU_2. The channel sensing times are as follows: $\tau_{PU2} > \tau_{PU1}$ and $\tau_{SU1} > \tau_{SU2}$.
- Explain the scope of IEEE Wireless Regional Area Networks.
- Differentiate between secondary and primary networks in the context of cognitive radio.
- Discuss the opportunistic nature of Mobile Relay Networks.
- Why the information sent in Machine-to-Machine communication is considerably smaller than the conventional data exchange?
- Explain the V2V2I mode of operation in the context of vehicular communication.
- Explain how two communication opportunities are separated by a disruption period in vehicular setups.
- What forces a communication opportunity to go away in V2V and V2I environments?
- What is the role of 3GPP in telecommunications?
- Name at least two design considerations that are important for LTE networks.
- Draw a labeled block diagram of LTE and its interconnection with 2G, 3G and IP networks.
- What is the difference between normal and extended cyclic prefixes used in LTE slots?
- Why TDD has seven frame configurations while FDD has only one?
- The special sub-frame is used in all TDD configurations. Why is it not used in the FDD frame structure?
- What do we understand by the "releases" of LTE network?
- How does CoMP help reduce inter-cell interference?
- What are the different types of carrier aggregation used in LTE-A?
- Give at least two reasons why relay nodes improve the network performance.
- Draw a labeled diagram showing the interconnection between EPC, eNB, DeBN and RNs. All interfaces must also be labeled.
- What are the peak data rates required by a technology to be categorized as 4G?
- What is the main motivation behind introducing 5G technology?
- What is mm-wave band?
- Why relaying provisions are mandatory for a 5G network?
- List at least three design considerations for UE Relaying concept.
- What is the role of D2D communication in realizing UE Relaying?
- Which type of UE Relaying is more relevant to the provision of ProSe?
- Explain how a D2D connection is setup between two DUEs.
- How is Wi-Fi Direct similar to cellular D2D communication?
- What is the significance of transmit power control and resource allocation in D2D communication.

Index

S. F. Hasan, *Emerging Trends in Communication Networks*,
SpringerBriefs in Electrical and Computer Engineering,
DOI: 10.1007/978-3-319-07389-7, © The Author(s) 2014